Spring Cloud
微服务开发实战
微课视频版

吴 胜 ◎ 编著

清华大学出版社

北京

内 容 简 介

本书由浅入深地介绍 Spring Cloud 微服务应用开发的基础知识。全书共 13 章，内容涵盖 Spring Cloud 概述；以及 Spring Cloud 路由、Spring Cloud 服务发现、Spring Cloud 认证与鉴权、Spring Cloud 断路器、Spring Cloud 配置中心、Spring Cloud 服务跟踪、Spring Cloud 消息、Spring Cloud 其他组件、Spring Cloud Alibaba、Dubbo 等应用；并以微课视频与案例相结合的方式实现手把手教学实战，重点讲解了 Spring Cloud 的综合应用案例以及 Service Mesh 与 Spring Cloud Sidecar。

本书内容通俗易懂，案例丰富，实战性强，适合 Spring Cloud 微服务应用开发的初学者(特别是在校学生)、企业级应用开发者和 Web 应用开发者；还可以作为高等学校相关专业的教材和自学参考书。

本书封面贴有清华大学出版社防伪标签，无标签者不得销售。
版权所有，侵权必究。举报: 010-62782989, beiqinquan@tup.tsinghua.edu.cn。

图书在版编目(CIP)数据

Spring Cloud 微服务开发实战: 微课视频版/吴胜编著. —北京: 清华大学出版社, 2020.1(2021.11重印)
(清华科技大讲堂)
ISBN 978-7-302-54234-6

Ⅰ. ①S… Ⅱ. ①吴… Ⅲ. ①互联网络—网络服务器 Ⅳ. ①TP368.5

中国版本图书馆 CIP 数据核字(2019)第 270780 号

责任编辑: 陈景辉　黄　芝
封面设计: 刘　键
责任校对: 胡伟民
责任印制: 杨　艳

出版发行: 清华大学出版社
网　　址: http://www.tup.com.cn, http://www.wqbook.com
地　　址: 北京清华大学学研大厦 A 座
邮　　编: 100084
社 总 机: 010-62770175
邮　　购: 010-62786544
投稿与读者服务: 010-62776969, c-service@tup.tsinghua.edu.cn
质量反馈: 010-62772015, zhiliang@tup.tsinghua.edu.cn
课件下载: http://www.tup.com.cn, 010-83470236

印 装 者: 三河市铭诚印务有限公司
经　　销: 全国新华书店
开　　本: 185mm×260mm
印　　张: 16.25
字　　数: 400 千字
版　　次: 2020 年 3 月第 1 版
印　　次: 2021 年 11 月第 2 次印刷
印　　数: 2501~3300
定　　价: 49.90 元

产品编号: 078485-01

前 言

正如 Martin Fowler 所指出的那样："微服务架构是一种将单一的应用程序划分为一组小型服务的方法,每个服务运行在自己的进程中,服务间通过轻量级的机制进行通信。"

目前,微服务架构尚未有统一的标准。但是,这丝毫不影响开发者对微服务架构的热情。认识上的差异使得大家采用不同方案实现微服务架构的应用,如 Dubbo 生态系统(Spring Cloud Alibaba 和 Dubbo 等)、Dropwizard、Spring Cloud 和 Service Mesh。

Spring Cloud 是众多功能的组合体,为微服务应用开发中遇到的众多问题提供了全面、优秀的"全家桶"式的解决方案。Spring Cloud 具有"开箱即用"、开发效率高、文档丰富、社区活跃等特点。因此,Spring Cloud 成为国内微服务开发领域的热门技术之一。

对于初学者而言,在进行 Spring Cloud 微服务应用开发时,相关技术略显复杂。比如,技术之间的关系、技术在整个微服务架构中的作用、项目的依赖管理和配置等问题都增加了 Spring Cloud 学习和开发的难度。Spring Cloud 以 Spring Boot 为基础,具备简单、易用的特点,而且 Spring Cloud 具有良好的开发工具、丰富的帮助文档,这些因素都有助于降低学习和开发难度。

考虑到初学者(特别是在校学生)往往对 Spring Boot、Spring Cloud、微服务的开发经验较少,为了降低 Spring Cloud 微服务开发的学习门槛,本书按照响应客户端服务请求的处理顺序和开发步骤的行文结构,安排与组织各个章节的内容。为了帮助读者更好地安排学习时间,同时也为了教师更好地安排教学,本书给出了各章的建议学时(建议学时分为理论学时和实践学时)。

章 内 容	建议理论学时	建议实践学时
第1章 Spring Cloud 概述	2	1
第2章 Spring Cloud 路由的应用	3	2
第3章 Spring Cloud 服务发现的应用	2	2
第4章 Spring Cloud 认证与鉴权的应用	2	2

续表

章　内　容	建议理论学时	建议实践学时
第 5 章 Spring Cloud 断路器的应用	2	1
第 6 章 Spring Cloud 配置中心的应用	2	1
第 7 章 Spring Cloud 服务跟踪的应用	1	1
第 8 章 Spring Cloud 消息的应用	1	1
第 9 章 Spring Cloud 其他组件的应用	2	1
第 10 章 Spring Cloud Alibaba 的应用	2	1
第 11 章 Dubbo 的应用	1	1
第 12 章 Spring Cloud 的综合应用案例	2	1
第 13 章 Service Mesh 与 Spring Cloud Sidecar	2	1
合计学时	24	16

在开设 Spring Cloud 微服务开发相关课程时，教师可以根据总课时、学生基础和教学目标等情况调整各章的学时。学习者也可以有选择地阅读各个章节的内容，并合理安排好学习时间。

为便于教学，本书配有教学视频、源代码、课件等配套资源。

（1）获取教学视频方式：读者可以先扫描本书封底的文泉云盘防盗码，再扫描书中相应的视频二维码，观看教学视频。

（2）获取源代码方式：先扫描本书封底的文泉云盘防盗码，再扫描下方二维码，即可获取。

（3）其他配套资源可以扫描本书封底的课件二维码下载。

关于如何直接使用源程序的说明

教材源程序

由于时间短，加上编者水平有限，书中难免有疏漏之处，敬请读者朋友批评指正。

作　者

2020 年 2 月

目 录

第1章　Spring Cloud 概述 ··· 1

1.1　Spring Cloud 简介 ··· 1

1.1.1　Spring 的构成 ··· 1

1.1.2　Spring Cloud 的构成 ··· 3

1.2　Spring Cloud 的特征 ··· 4

1.2.1　Spring Boot 与 Spring Cloud 的共同特点 ··· 4

1.2.2　Spring Cloud 的其他特点 ··· 5

1.3　配置开发环境 ·· 5

1.3.1　安装 JDK ·· 5

1.3.2　安装 IntelliJ IDEA ··· 5

1.4　创建项目与实现微服务 ··· 6

1.4.1　利用 IDEA 创建项目 ··· 6

1.4.2　创建项目的基本构成情况 ··· 10

1.4.3　基于 Spring Boot 的微服务实现 ··· 11

1.5　Spring Cloud 微服务开发起步 ··· 12

1.5.1　软件生命周期 ··· 12

1.5.2　Spring Cloud 响应服务请求的处理顺序 ··· 13

1.5.3　Spring Cloud 微服务开发的步骤 ··· 13

习题 1 ·· 14

第2章　Spring Cloud 路由的应用 ·· 15

2.1　服务网关与 Spring Cloud 路由简介 ··· 15

2.1.1　服务网关的作用 ··· 15

2.1.2　Spring Cloud 路由的相关技术 ·· 16

2.2　Spring Cloud Gateway 路由的应用 ·· 17

2.2.1 创建项目并添加依赖 …… 17
2.2.2 创建类 HelloController …… 17
2.2.3 修改配置文件 application.properties …… 18
2.2.4 修改入口类 …… 18
2.2.5 运行程序 …… 19
2.2.6 程序扩展 …… 19
2.3 Spring Cloud Gateway 过滤器的应用 …… 20
2.3.1 创建项目并添加依赖 …… 20
2.3.2 创建类 HelloController …… 20
2.3.3 创建类 ElapsedFilter …… 20
2.3.4 修改入口类 …… 21
2.3.5 运行程序 …… 22
2.4 Spring Cloud Feign 的应用 …… 22
2.4.1 创建项目并添加依赖 …… 23
2.4.2 创建接口 FeignService …… 23
2.4.3 创建类 FeignController …… 23
2.4.4 修改入口类 …… 24
2.4.5 运行程序 …… 24
2.5 Spring Cloud Ribbon 的应用 …… 25
2.5.1 创建项目并添加依赖 …… 25
2.5.2 创建类 HelloController …… 26
2.5.3 修改配置文件 application.properties …… 26
2.5.4 修改入口类 …… 26
2.5.5 运行程序 …… 27
2.5.6 程序扩展 …… 27
2.6 Spring Cloud Zuul 的应用 …… 29
2.6.1 创建项目并添加依赖 …… 29
2.6.2 创建配置文件 application.yml …… 29
2.6.3 修改入口类 …… 30
2.6.4 运行程序 …… 30

| | | 2.6.5 | 程序扩展 ……………………………………………………… | 30 |

习题 2 ……………………………………………………………………………… 31

第 3 章　Spring Cloud 服务发现的应用 …………………………………………… 32

3.1　Spring Cloud 服务注册与发现的简介 ……………………………………… 32

- 3.1.1　服务的注册和发现 ……………………………………………… 32
- 3.1.2　Spring Cloud 服务发现解决方案 ……………………………… 33

3.2　Spring Cloud Eureka 的应用 ……………………………………………… 33

- 3.2.1　Spring Cloud Eureka 注册中心的实现 ………………………… 34
- 3.2.2　Spring Cloud Eureka 服务提供者的实现 ……………………… 35
- 3.2.3　Spring Cloud Eureka 服务消费者的实现 ……………………… 38
- 3.2.4　运行程序 ………………………………………………………… 40

3.3　Spring Cloud Consul 的应用 ……………………………………………… 42

- 3.3.1　Spring Cloud Consul 服务提供者的实现 ……………………… 42
- 3.3.2　Spring Cloud Consul 服务消费者的实现 ……………………… 44
- 3.3.3　运行程序 ………………………………………………………… 47

3.4　Spring Cloud Zookeeper 的应用 …………………………………………… 48

- 3.4.1　Spring Cloud Zookeeper 服务提供者的实现 ………………… 48
- 3.4.2　Spring Cloud Zookeeper 服务消费者的实现 ………………… 49
- 3.4.3　运行程序 ………………………………………………………… 51

习题 3 ……………………………………………………………………………… 53

第 4 章　Spring Cloud 认证与鉴权的应用 ………………………………………… 54

4.1　Spring Cloud Security 的简单应用 ………………………………………… 54

- 4.1.1　创建项目并添加依赖 …………………………………………… 54
- 4.1.2　创建类 HelloController ………………………………………… 55
- 4.1.3　创建配置文件 application.yml ………………………………… 55
- 4.1.4　运行程序 ………………………………………………………… 55
- 4.1.5　程序扩展 ………………………………………………………… 56

4.2　Spring Cloud OAuth 2 的简单应用 ………………………………………… 58

- 4.2.1　创建项目并添加依赖 …………………………………………… 58
- 4.2.2　创建类 HelloController ………………………………………… 58

4.2.3　创建文件 index.html ………………………………………………… 59
		4.2.4　创建配置文件 application.yml ……………………………………… 59
		4.2.5　修改入口类 ……………………………………………………………… 60
		4.2.6　运行程序 ………………………………………………………………… 60
	4.3　JWT 的简单应用 ………………………………………………………………… 61
		4.3.1　创建项目并添加依赖 …………………………………………………… 61
		4.3.2　创建类 User ……………………………………………………………… 62
		4.3.3　创建类 TokenUserAuthentication ……………………………………… 63
		4.3.4　创建类 JwtUtil …………………………………………………………… 64
		4.3.5　创建类 HelloController ………………………………………………… 65
		4.3.6　创建文件 index.html …………………………………………………… 67
		4.3.7　创建配置文件 application.yml ……………………………………… 68
		4.3.8　修改入口类 ……………………………………………………………… 69
		4.3.9　运行程序 ………………………………………………………………… 69
	4.4　Gateway、JWT、Actuator 的综合应用 ………………………………………… 71
		4.4.1　创建项目并添加依赖 …………………………………………………… 71
		4.4.2　创建类 JwtUtil …………………………………………………………… 71
		4.4.3　创建类 HelloController ………………………………………………… 72
		4.4.4　创建配置文件 application.yml ……………………………………… 73
		4.4.5　修改入口类 ……………………………………………………………… 73
		4.4.6　运行程序 ………………………………………………………………… 74
	4.5　Eureka、Zuul、OAuth2 和 JWT 的综合应用 ………………………………… 75
		4.5.1　zuul-server 的实现 ……………………………………………………… 75
		4.5.2　auth-server 的实现 ……………………………………………………… 77
		4.5.3　client-a 的实现 …………………………………………………………… 80
		4.5.4　运行程序 ………………………………………………………………… 84
习题 4 ………………………………………………………………………………………… 86
第 5 章　Spring Cloud 断路器的应用 …………………………………………………… 87
	5.1　Spring Cloud Hystrix 的应用 …………………………………………………… 87
		5.1.1　创建项目并添加依赖 …………………………………………………… 88

        5.1.2 创建接口 HiService ……………………………………… 88
        5.1.3 创建类 HiController ……………………………………… 89
        5.1.4 创建类 HelloController …………………………………… 89
        5.1.5 修改配置文件 application.properties …………………… 90
        5.1.6 修改入口类 ………………………………………………… 90
        5.1.7 运行程序 …………………………………………………… 91
    5.2 Spring Cloud Hystrix Dashboard 的应用 ………………………… 92
        5.2.1 添加依赖 …………………………………………………… 92
        5.2.2 修改入口类 ………………………………………………… 92
        5.2.3 运行程序 …………………………………………………… 93
    5.3 Spring Cloud Turbine 的应用 ……………………………………… 95
        5.3.1 创建项目并添加依赖 ……………………………………… 95
        5.3.2 修改配置文件 application.properties …………………… 95
        5.3.3 修改入口类 ………………………………………………… 96
        5.3.4 运行程序 …………………………………………………… 96
    习题 5 ……………………………………………………………………… 97
第 6 章 Spring Cloud 配置中心的应用 ……………………………………… 98
    6.1 Spring Cloud Config Server 的应用 ……………………………… 98
        6.1.1 创建项目并添加依赖 ……………………………………… 99
        6.1.2 修改配置文件 application.properties …………………… 99
        6.1.3 修改入口类 ………………………………………………… 99
        6.1.4 运行程序 …………………………………………………… 100
    6.2 Spring Cloud Config Client 的应用 ……………………………… 100
        6.2.1 创建项目并添加依赖 ……………………………………… 101
        6.2.2 创建类 HelloController …………………………………… 101
        6.2.3 修改配置文件 application.properties …………………… 102
        6.2.4 运行程序 …………………………………………………… 102
    6.3 Spring Cloud Consul 的应用 ……………………………………… 102
        6.3.1 创建项目并添加依赖 ……………………………………… 102
        6.3.2 创建配置文件 application.yml …………………………… 103

6.3.3　创建配置文件 bootstrap.yml ………………………………………… 103
　　6.3.4　修改入口类 ………………………………………………………… 104
　　6.3.5　运行程序 …………………………………………………………… 105
6.4　Spring Cloud Zookeeper 的应用 ……………………………………………… 107
　　6.4.1　创建项目并添加依赖 ……………………………………………… 107
　　6.4.2　创建类 HelloController …………………………………………… 107
　　6.4.3　创建配置文件 bootstrap.yml ………………………………………… 108
　　6.4.4　运行程序 …………………………………………………………… 108
习题 6 …………………………………………………………………………………… 109

第 7 章　Spring Cloud 服务跟踪的应用 …………………………………………… 110

7.1　Spring Cloud Sleuth 的应用 …………………………………………………… 110
　　7.1.1　创建项目并添加依赖 ……………………………………………… 110
　　7.1.2　创建类 SleuthService ………………………………………………… 111
　　7.1.3　创建类 SchedulingService …………………………………………… 111
　　7.1.4　创建类 ThreadConfig ………………………………………………… 112
　　7.1.5　创建类 HelloController ……………………………………………… 113
　　7.1.6　修改配置文件 application.properties ………………………………… 114
　　7.1.7　运行程序 …………………………………………………………… 114
7.2　Spring Cloud Zipkin 的应用 …………………………………………………… 116
　　7.2.1　创建项目 zipkinexample ……………………………………………… 116
　　7.2.2　创建项目 zipkinclient1 ……………………………………………… 117
　　7.2.3　创建项目 zipkinuser1 ………………………………………………… 119
　　7.2.4　运行程序 …………………………………………………………… 120
习题 7 …………………………………………………………………………………… 121

第 8 章　Spring Cloud 消息的应用 ………………………………………………… 122

8.1　Spring Cloud Bus 的应用 ……………………………………………………… 122
　　8.1.1　Spring Cloud Config Server 的应用 …………………………………… 122
　　8.1.2　Spring Cloud Bus 的应用实现 ………………………………………… 124
　　8.1.3　运行程序 …………………………………………………………… 125
8.2　Spring Cloud Stream 的应用 …………………………………………………… 127

8.2.1　创建项目并添加依赖 …………………… 127
　　　8.2.2　创建接口 Sink …………………… 127
　　　8.2.3　创建类 SinkReceiver …………………… 128
　　　8.2.4　创建配置文件 application.yml …………………… 128
　　　8.2.5　运行程序 …………………… 128
　习题 8 …………………… 129

第 9 章　Spring Cloud 其他组件的应用 …………………… 130

9.1　Spring Cloud Task 的应用 …………………… 130
　　　9.1.1　创建项目并添加依赖 …………………… 130
　　　9.1.2　创建类 ScheduledTask …………………… 131
　　　9.1.3　创建类 HelloController …………………… 131
　　　9.1.4　创建配置文件 application.yml …………………… 131
　　　9.1.5　修改入口类 …………………… 132
　　　9.1.6　运行程序 …………………… 132
9.2　Spring Cloud Function 的应用 …………………… 133
　　　9.2.1　创建项目并添加依赖 …………………… 133
　　　9.2.2　创建类 Greeter …………………… 133
　　　9.2.3　创建类 HelloController …………………… 133
　　　9.2.4　运行程序 …………………… 134
9.3　Cloud Foundry 的应用 …………………… 134
　　　9.3.1　Cloud Foundry 简介 …………………… 134
　　　9.3.2　利用 Cloud Foundry 平台部署 Spring Boot 应用 …………………… 134
　习题 9 …………………… 135

第 10 章　Spring Cloud Alibaba 的应用 …………………… 136

10.1　Spring Cloud Alibaba 简介 …………………… 136
　　　10.1.1　Spring Cloud Alibaba 主要功能 …………………… 136
　　　10.1.2　Spring Cloud Alibaba 组件 …………………… 137
10.2　Nacos Config 的应用 …………………… 138
　　　10.2.1　创建项目并添加依赖 …………………… 138
　　　10.2.2　创建类 ConfigController …………………… 138

10.2.3 创建并修改配置文件 bootstrap.properties ······ 139
10.2.4 运行程序 ······ 139
10.3 Nacos Discovery 的应用 ······ 141
10.3.1 服务提供者的实现 ······ 141
10.3.2 服务消费者的实现 ······ 142
10.3.3 运行程序 ······ 144
10.4 Sentinel 的应用 ······ 144
10.4.1 创建项目并添加依赖 ······ 145
10.4.2 创建类 HelloController ······ 145
10.4.3 修改配置文件 application.properties ······ 146
10.4.4 运行程序 ······ 146
10.5 ACM 的应用 ······ 147
10.5.1 辅助工作 ······ 148
10.5.2 创建项目并添加依赖 ······ 148
10.5.3 创建类 SampleController ······ 149
10.5.4 修改配置文件 application.properties ······ 149
10.5.5 运行程序 ······ 149
习题 10 ······ 150

第 11 章 Dubbo 的应用 ······ 151

11.1 Dubbo 简介 ······ 151
11.1.1 Dubbo 主要功能 ······ 151
11.1.2 Dubbo Spring Boot 简介 ······ 152
11.2 Dubbo 的简单应用 ······ 152
11.2.1 服务提供者的实现 ······ 152
11.2.2 服务消费者的实现 ······ 155
11.2.3 运行程序 ······ 156
11.3 Dubbo Spring Boot 的应用 ······ 157
11.3.1 服务提供者的实现 ······ 157
11.3.2 服务消费者的实现 ······ 159
11.3.3 运行程序 ······ 160

11.4 Spring Cloud Dubbo 的应用 ……………………………………………… 160
11.4.1 服务提供者的实现 ……………………………………………… 160
11.4.2 服务消费者的实现 ……………………………………………… 162
11.4.3 运行程序 …………………………………………………………… 163
习题 11 ……………………………………………………………………………… 164

第 12 章 Spring Cloud 的综合应用案例 ……………………………………… 165
12.1 实现配置中心 case-config-server ……………………………………… 165
12.1.1 创建项目并添加依赖 …………………………………………… 165
12.1.2 创建配置文件 application.yml ………………………………… 165
12.1.3 修改入口类 ……………………………………………………… 166
12.1.4 运行程序 ………………………………………………………… 166
12.2 实现客户端服务 case-eureka-user-client ……………………………… 167
12.2.1 创建项目并添加依赖 …………………………………………… 167
12.2.2 创建类 User ……………………………………………………… 168
12.2.3 创建接口 UserDao ……………………………………………… 169
12.2.4 创建类 UserController ………………………………………… 169
12.2.5 修改和创建配置文件 …………………………………………… 170
12.2.6 修改入口类 ……………………………………………………… 171
12.2.7 运行程序 ………………………………………………………… 172
12.3 实现服务消费端 case-user-ribbon ……………………………………… 173
12.3.1 创建项目并添加依赖 …………………………………………… 173
12.3.2 创建类 User ……………………………………………………… 173
12.3.3 创建类 UserRibbonService …………………………………… 174
12.3.4 创建类 UserController ………………………………………… 175
12.3.5 创建配置文件 application.yml ………………………………… 176
12.3.6 修改入口类 ……………………………………………………… 176
12.3.7 运行程序 ………………………………………………………… 177
12.4 实现服务消费端 case-service …………………………………………… 177
12.4.1 创建项目并添加依赖 …………………………………………… 177
12.4.2 创建类 User ……………………………………………………… 178

12.4.3 创建接口 UserFeignService ……………… 178
12.4.4 创建类 UserController ……………… 178
12.4.5 修改配置文件 application.properties ……………… 179
12.4.6 修改入口类 ……………… 179
12.4.7 运行程序 ……………… 180
12.5 实现路由网关 case-zuul ……………… 180
12.5.1 创建项目并添加依赖 ……………… 180
12.5.2 创建配置文件 application.yml ……………… 181
12.5.3 修改入口类 ……………… 182
12.5.4 运行程序 ……………… 182

习题 12 ……………… 183

第13章 Service Mesh 与 Spring Cloud Sidecar ……………… 184

13.1 Service Mesh 概述 ……………… 184
13.1.1 Service Mesh 简介 ……………… 184
13.1.2 Service Mesh 的特点 ……………… 186
13.1.3 数据面和控制面 ……………… 186
13.2 Linkerd 和 Envoy 简介 ……………… 187
13.2.1 Linkerd 简介 ……………… 187
13.2.2 Envoy 简介 ……………… 187
13.3 Istio 概述 ……………… 188
13.3.1 Istio 简介 ……………… 188
13.3.2 Istio 核心功能 ……………… 189
13.3.3 Istio 架构 ……………… 190
13.3.4 Istio 应用的模拟 ……………… 191
13.4 Conduit 概述 ……………… 192
13.4.1 Conduit 简介 ……………… 192
13.4.2 Conduit 架构 ……………… 192
13.5 国内 Service Mesh 实践简介 ……………… 193
13.5.1 SOFAMesh 简介 ……………… 193
13.5.2 Dubbo Mesh 简介 ……………… 193

- 13.5.3 华为服务网格简介 …… 194
- 13.5.4 京东服务网格简介 …… 195
- 13.5.5 新浪微博 Weibo Mesh 简介 …… 195
- 13.5.6 云帮 Rainbond 服务网格简介 …… 196
- 13.6 Spring Cloud Sidecar 的应用 …… 196
  - 13.6.1 创建项目并添加依赖 …… 197
  - 13.6.2 修改配置文件 application.properties …… 197
  - 13.6.3 修改入口类 …… 197
  - 13.6.4 创建 node-service.js …… 198
  - 13.6.5 运行程序 …… 198
- 习题 13 …… 200

附录 A 相关软件的安装和配置 …… 201
- A.1 JDK 的安装和配置 …… 201
- A.2 Consul 的配置 …… 202
- A.3 ZooKeeper 的配置 …… 204
- A.4 Nacos 服务器的配置 …… 206

附录 B 基于 Feign 实现文件传送 …… 207
- B.1 实现 Eureka 服务器项目 mweathereurekaserver …… 207
  - B.1.1 新建项目并添加依赖 …… 207
  - B.1.2 修改入口类与创建配置文件 …… 207
- B.2 实现文件接收者项目 feignserver …… 208
  - B.2.1 新建项目并添加依赖 …… 208
  - B.2.2 修改入口类和配置文件 …… 208
- B.3 实现文件传送者项目 feignclient …… 208
  - B.3.1 新建项目并添加依赖 …… 208
  - B.3.2 创建接口、类和修改配置文件 …… 208
- B.4 程序运行结果 …… 208

附录 C 基于 Ribbon 实现文件上传 …… 211
- C.1 实现文件上传服务提供者项目 uploadfile …… 211
  - C.1.1 新建项目并添加依赖 …… 211

　　　　C.1.2　创建类 ··· 211
　　　　C.1.3　新建文件和修改配置文件 ·· 212
　　C.2　实现文件上传服务消费者项目 fileuser ······································ 212
　　　　C.2.1　新建项目并添加依赖 ·· 212
　　　　C.2.2　创建类、修改配置文件和配置文件 ···························· 212
　　C.3　程序运行结果 ·· 212

附录D　简易天气预报系统的实现 ··· 215
　　D.1　实现天气服务提供者项目 weatherbasic ····································· 215
　　　　D.1.1　新建项目并添加依赖 ·· 215
　　　　D.1.2　创建类、接口并修改配置文件 ·································· 215
　　D.2　实现天气服务消费者项目 weatherclient ····································· 216
　　　　D.2.1　新建项目并添加依赖 ·· 216
　　　　D.2.2　创建类 ··· 216
　　　　D.2.3　新建文件和修改配置文件 ·· 216
　　D.3　程序运行结果 ·· 216

附录E　Apollo 和 Zuul 的整合开发 ·· 219
　　E.1　Apollo 配置中心的准备和启动 ·· 219
　　　　E.1.1　Apollo 配置中心的准备 ··· 219
　　　　E.1.2　Apollo 配置中心的启动 ··· 220
　　E.2　本案例的结构说明和 Apollo 配置中心的内容设置 ························· 222
　　　　E.2.1　本案例的结构说明 ··· 222
　　　　E.2.2　Apollo 配置中心的内容设置 ····································· 223
　　E.3　实现服务提供者项目 apolloconfig ··· 224
　　　　E.3.1　新建项目并添加依赖 ·· 224
　　　　E.3.2　创建类、文件和修改配置文件 ··································· 225
　　E.4　实现服务提供者项目 apollouser ·· 225
　　　　E.4.1　新建项目并添加依赖 ·· 225
　　　　E.4.2　创建类、文件和修改配置文件 ··································· 225
　　E.5　实现 zuul 项目 zuulapollo ·· 225
　　　　E.5.1　新建项目并添加依赖 ·· 225

		E.5.2	创建类、修改入口类和配置文件 ………………………	225

E.6 程序运行结果 …………………………………………………………………… 226
  E.6.1 apolloconfig 服务运行结果 ……………………………………… 226
  E.6.2 apollouser 服务运行结果 ………………………………………… 226
  E.6.3 zuulapollo 服务运行结果 ………………………………………… 227

附录 F　Spring Cloud 在微信小程序的简单应用 …………………………… 228
F.1 启动作为后台的 Spring Cloud 微服务 ………………………………………… 228
  F.1.1 启动 Apollo 配置中心 …………………………………………… 228
  F.1.2 保持后台服务不变 ………………………………………………… 229
  F.1.3 在浏览器中直接访问微服务的结果 ……………………………… 229
F.2 前端微信小程序应用的实现 ……………………………………………………… 229
  F.2.1 微信小程序的开发工具安装和项目创建 ………………………… 229
  F.2.2 创建项目并新建、修改文件 ……………………………………… 230
  F.2.3 微信小程序项目的运行结果 ……………………………………… 230
F.3 Spring Cloud 微服务和微信小程序整合的关键点 …………………………… 232
  F.3.1 两者关联的关键代码 ……………………………………………… 232
  F.3.2 注意事项 …………………………………………………………… 232

附录 G　Spring Cloud 和 Vue.js 的整合开发 ……………………………… 234
G.1 在 IDEA 中整合 Spring Cloud 和 Vue.js ……………………………………… 234
  G.1.1 Vue.js 的安装 ……………………………………………………… 234
  G.1.2 在 IDEA 中集成 Vue.js …………………………………………… 236
G.2 Spring Cloud 微服务和 Vue.js 整合示例的实现 ……………………………… 238
  G.2.1 创建 Vue.js 项目 helloworld ……………………………………… 238
  G.2.2 后台服务 …………………………………………………………… 239
  G.2.3 运行结果 …………………………………………………………… 239
G.3 Spring Cloud 微服务和 Vue.js 整合的关键点 ………………………………… 240
  G.3.1 两者整合的关键 …………………………………………………… 240
  G.3.2 结果对比 …………………………………………………………… 241

参考文献 …………………………………………………………………………………… 242

第 1 章

Spring Cloud 概述

本章介绍 Spring Cloud 的简介,然后介绍 Spring Cloud 的特征,如何配置 Spring Cloud 开发环境,如何创建项目与实现微服务,Spring Cloud 微服务开发的起步等内容。

1.1 Spring Cloud 简介

本节通过介绍 Spring 的构成,来说明 Spring Cloud 在整个 Spring 框架中的位置。在此基础上,介绍 Spring Cloud 的子项目。

1.1.1 Spring 的构成

Spring 作为 EJB 的颠覆者,因其轻量级的开发方式而很快地被业界接受。Spring 提供了非常多的、值得注意的子项目;了解这些子项目可以帮助读者更好地理解 Spring 设计架构、思想并学会使用 Spring。Spring 的整个生态系统包括如下内容。

Spring Framework Core:是 Spring 的核心项目,其中包含了一系列 IoC 容器的设计,提供了依赖注入的实现;还集成了面向切面编程(AOP)、Spring MVC、JDBC、事务处理模块。

Spring Boot：提供了能够快速构建 Spring 应用的方法，实现了"开箱即用"；使用默认的 Java 配置来实现快速开发，并实现"即时运行"。

Spring Cloud：提供了用于快速构建分布式系统（微服务）的工具集，利用它进行开发时，往往需要基于 Spring Boot。

Spring Batch：提供了用于构建批处理应用和自动化操作的框架，专门用于离线分析程序、数据批处理等场景。

Spring Data：对主流的关系型数据库提供支持；并提供使用非关系型数据的功能，例如，数据存储在非关系型数据库或 Map-Reduce 的分布式存储环境和云计算存储环境等。

Spring Security：通过用户认证、授权、安全服务等工具保护应用。

Spring Security OAuth：OAuth 是一个第三方的模块，提供一个开放的协议，通过这个协议，前端应用可以对 Web 应用进行简单而标准的安全调用。

Spring Web Flow：基于 Spring MVC 进行 Web 应用开发。它是 Web 工作流引擎，定义了一种特定的语言来描述工作流；同时高级的工作流控制器引擎可以管理会话状态。

Spring BlazeDS Integration：用于提供 Spring 与 Adobe Flex 技术集成的模块。

Spring Dynamic Modules：用于提供 Spring 运行在 OSGi 平台上面向 Java 的动态模型系统。

Spring Integration：通过消息机制为企业的数据集成提供解决方案。

Spring AMQP：高级消息队列协议（Advanced Message Queuing Protocol，AMQP）支持 Java 和 .NET 两个版本。AMQP 是一个用于提供统一消息服务的应用层标准协议，是一个开放标准，为面向消息的中间件（如 RabbitMQ 等）而设计。

Spring .NET：为 .NET 提供 Spring 相关的技术支持，如 IoC 容器、AOP 等。

Spring for Android：为 Android 终端应用开发提供 Spring 支持。

Spring Mobile：为移动终端的服务器应用开发提供支持。

Spring Social：是 Spring 框架的扩展，用于提供与社交网 SNS 服务 API（如 FaceBook、新浪微博和 Twitter 等）的集成。

Spring XD：用于简化大数据应用和开发。

Spring HATEOAS：基于 HATEOAS 原则简化 REST 服务开发流程。超文本驱动（Hypermedia As The Engine Of Application State，HATEOAS），又称为将超媒

体作为应用状态的引擎。

Spring Web Services：是基于 Spring 的 Web 服务框架，提供 SOAP 服务开发。SOAP 是简单对象访问协议（Simple Object Access Protocol）的英文缩写。

Spring LDAP：用于简化使用轻量目录访问协议（Lightweight Directory Access Protocol，LDAP）开发。

Spring Session：提供 API，并管理用户会话信息。

1.1.2 Spring Cloud 的构成

Spring Cloud 作为微服务开发的工具集，包括在分布式系统开发时要用到的许多子项目。Spring Cloud 的子项目如下所述。

Spring Cloud Config：配置管理开发工具包，目前支持本地存储、远程仓库（如 Git）存储配置信息。

Spring Cloud Bus：事件、消息总线，用于在集群中传播状态变化，可与 Spring Cloud Config 联合，实现热部署。

Spring Cloud Netflix：针对多种 Netflix 组件提供的开发工具包，其中包括 Eureka、Hystrix、Zuul 等。

Spring Cloud for Cloud Foundry：通过 OAuth2 协议绑定服务到 Cloud Foundry，Cloud Foundry 是由 VMware 推出的开源 PaaS 云平台。

Spring Cloud Sleuth：日志收集工具包，封装了 Dapper、Zipkin 和 HTrace 等。

Spring Cloud Data Flow：大数据操作工具，通过命令行方式操作数据流。

Spring Cloud Security：安全工具包，为应用程序添加安全控制。

Spring Cloud Consul：封装了 Consul 操作，可以与 Docker 容器无缝集成。

Spring Cloud Zookeeper：操作 Apache ZooKeeper 的工具包。Apache ZooKeeper 是 Apache 的一个软件项目，它为大型分布式计算提供开源的分布式配置、同步服务和命名注册。Spring Cloud Zookeeper 是对 Apache ZooKeeper 的应用。后面章节用 ZooKeeper 指代 Apache ZooKeeper，用 Zookeeper 指代 Spring Cloud Zookeeper。相比而言，Zookeeper 比 ZooKeeper 的粒度（规模）要小。

Spring Cloud Stream：数据流操作开发包，封装了与 Redis、RabbitMQ 等数据库、消息代理的收发消息机制。

Spring Cloud CLI：利用 Spring Boot CLI 可以以命令行方式，快速地建立云组件。

本书开始编写时，Spring Cloud 最新版本为 Finchley.RELEASE 版；本书的实例主要基于此版本编写的，Finchley.RELEASE 版的 Spring Cloud 是基于 2.0.x 版 Spring Boot 的。

1.2 Spring Cloud 的特征

Spring Cloud 开发以 Spring Boot 为基础，Spring Cloud 开发步骤和方法与 Spring Boot 的开发步骤和方法相同。因此，Spring Cloud 具备 Spring Boot 的特点。

1.2.1 Spring Boot 与 Spring Cloud 的共同特点

(1) 约定大于配置。通过代码结构、注解的约定和命名规范等方式来减少配置，并采用更加简洁的配置方式来替代 XML 配置；减少冗余代码和强制的 XML 配置。

(2) 能创建基于 Spring 框架的独立应用程序。

(3) 内嵌有 Tomcat(或 Netty)，打包方式不再强制要求打包成 War 包来部署，可以直接采用 Jar 包。

(4) 简化 Maven 配置，并推荐使用 Gradle 替代 Maven 进行项目管理。Maven 用于项目的构建，主要可以对依赖包进行管理，Maven 将项目所使用的依赖包信息放到 pom.xml 文件的<dependencies></dependencies>节点之间。

(5) 能自动配置 Spring。

(6) 提供生产就绪型功能。即提供了一些大型项目中常见的非功能特性，如嵌入式服务器、安全、指标、健康检测、外部配置等内容。

(7) 定制"开箱即用"的 Starter，没有代码生成，也无须 XML 配置；还可以修改默认值以满足特定的需求。

(8) 为 Spring 开发者提供更快速的入门体验。Spring Boot 不是对 Spring 进行功能上的增强，而是提供了一种更快速的 Spring 使用方法。

(9) 对主流框架无配置集成，自动整合第三方框架，如 Struts。

(10) 使用注解使编码变得更加简单。

(11) Spring Boot 可与基于 Spring Cloud 的微服务开发无缝结合。

1.2.2　Spring Cloud 的其他特点

Spring Cloud 除了继承了 Spring Boot 的特点外,还具备一些其他特点。

(1) Spring Cloud 组件丰富,功能齐全,适用的环境多,为微服务架构实现提供了较完美的"全家桶"式解决方案。

(2) 基于成熟的轻量级组件。Spring Cloud 的组件往往是对已有的、成熟的组件的封装。Spring Cloud 所封装的组件大多是轻量级的组件,为微服务的实现提供了保障。

(3) 选型灵活、中立。Spring Cloud 组件之间的关系相对独立,并且往往可以对同一问题提供多个解决方案,开发人员可以根据需要选择不同的技术组合来实现微服务开发。

1.3　配置开发环境

配置 Spring Cloud 的开发环境,需要先安装 JDK;然后选择并安装一款合适的开发工具。本书以 IntelliJ IDEA 作为开发工具。

1.3.1　安装 JDK

JDK 的安装步骤详见附录 A。完成安装后,打开 Windows 命令处理程序 CMD,执行命令 java -version,如果见到如图 1-1 所示的版本信息就说明 JDK 已经安装成功。

图 1-1　JDK 安装成功后显示的版本信息

1.3.2　安装 IntelliJ IDEA

可以从 IntelliJ IDEA(以下简称为 IDEA)官网(https://www.jetbrains.com)下

载免费的社区版或者旗舰试用版 IDEA，然后进行安装。安装完成后打开 IDEA，将显示如图 1-2 所示的欢迎界面。

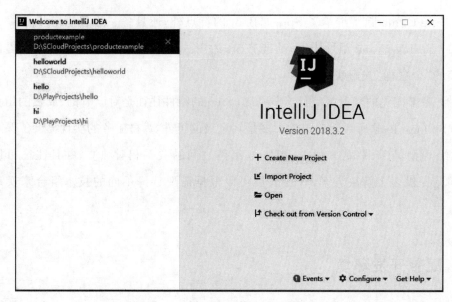

图 1-2　IDEA 启动后的欢迎界面

1.4　创建项目与实现微服务

视频讲解

可以用不同的工具创建 Spring Cloud 项目，本节介绍利用 IDEA 创建项目的方法。

1.4.1　利用 IDEA 创建项目

先在如图 1-2 所示的欢迎界面中选择 Create New Project 超链接，进入创建新项目（New Project）界面；并选择 Spring Initializr 类型的项目，如图 1-3 所示。

接着，单击 Next 按钮跳转到项目信息的设置界面，IDEA 创建新项目时要根据项目情况设置项目的元数据（Project Metadata）；设置项目元数据的界面如图 1-4 所示。

在所创建项目 Group 文本框中输入 com.bookcode，在 Artifact 文本框中输入 springcloud-helloworlds，如图 1-4 所示。在用于创建项目的管理工具类型 Type 的下拉列表框中选择 Maven Project。目前 Maven 的参考资料比 Gradle 的参考资料多且更容易获得，所以本书使用 Maven，进行项目管理。在开发语言 Language 的下拉列

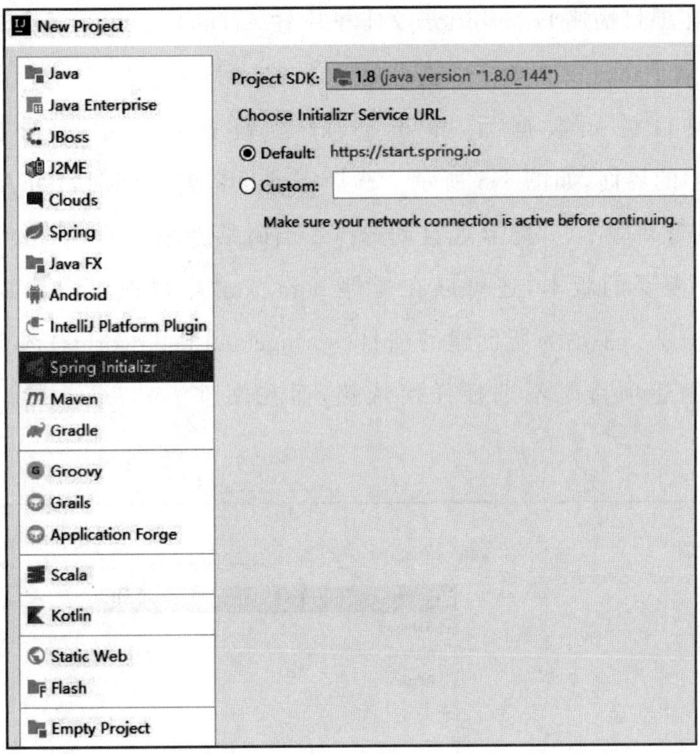

图 1-3　IDEA 中创建 New Project 时选择 Spring Initializr 类型项目的界面

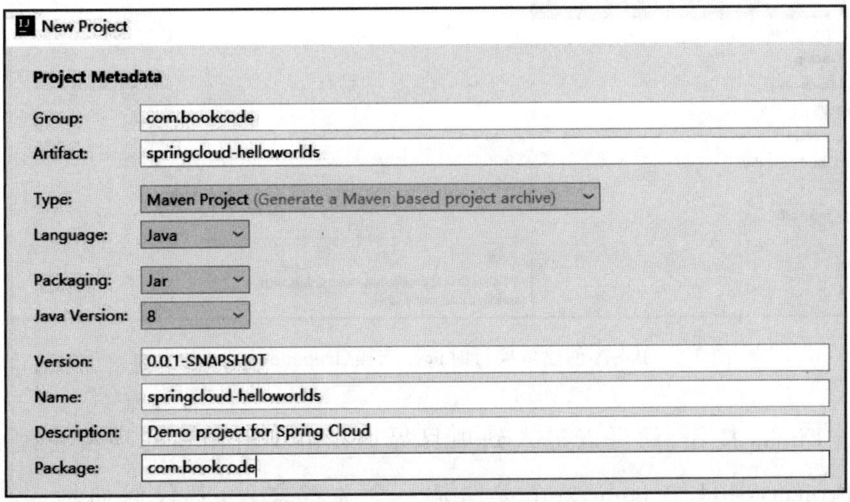

图 1-4　IDEA 创建新项目时设置项目元数据（Project Metadata）的界面

表框中选择 Java；在打包方式 Packaging 的下拉列表框中选择 Jar；在 Java Version 的下拉列表框中选择 8（也称为 1.8）；所创建项目的版本 Version 文本框中保留自动生成的 0.0.1-SNAPSHOT；在项目名称 Name 文本框中保留自动生成的 springcloud-helloworlds；在项目描述 Description 文本框中输入 Demo project for Spring Cloud；在项目默认包名 Package 文本框中输入 com.bookcode。

　　填写完项目的元数据后，单击 Next 按钮就可以进入选择项目依赖（Dependencies）的界面，如图 1-5 所示。选择完项目依赖的同时，IDEA 会自动选择 Spring Boot 的最新版本（并隐含地自动选择了当时最新版本的 Spring Cloud），也可以手动选择所需要的版本，还可以在文件 pom.xml 中修改 Spring Boot 和 Spring Cloud 的版本信息。手动为所创建的项目（springcloud-helloworlds）选择 Gateway 依赖，也可以在创建项目时不选择任何依赖，而在文件 pom.xml 中添加所需要的依赖。

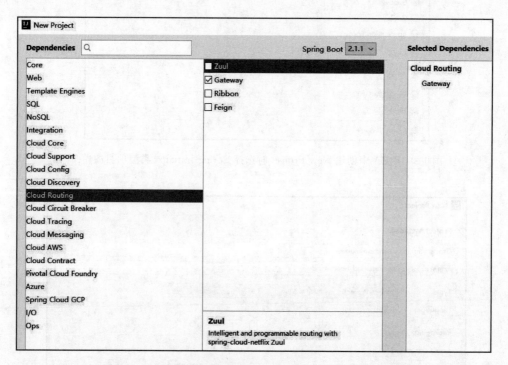

图 1-5　IDEA 创建新项目时选择依赖（Dependencies）的界面

　　单击 Finish 按钮，就可以进入到项目界面。由于所创建的项目管理类型为 Maven Project，项目中 pom.xml 文件是一个关键文件，其代码如例 1-1 所示。<parent></parent>之间的内容表示父依赖，是一般项目都要用到的基础内容，其中包含了项目中用到的 Spring Boot 的版本信息。<properties></properties>之间的内容表示了项目中所用到的编码格式、Java 版本和 Spring Cloud 版本信息。由于本书选择稳定的 Spring Cloud 版本（Finchley.RELEASE 版），Spring Boot 采用 2.1.1.RELEASE 版。

因此,需要修改例 1-1 中 pom.xml 文件的 Spring Boot 和 Spring Cloud 版本信息。<dependencies></dependencies>之间的内容包含了项目所要用到的依赖信息;<dependencyManagement></dependencyManagement>之间的内容表示了对 Spring Cloud 的依赖管理。<build></build>之间的内容表示了编译运行时要用到的相关插件。<repositories></repositories>之间的内容表示仓库信息。关于 Maven 依赖的更多情况,将在后面的示例中介绍。

【例 1-1】 pom.xml 文件代码示例。

```xml
<?xml version = "1.0" encoding = "UTF-8"?>
<project xmlns = "http://maven.apache.org/POM/4.0.0" xmlns:xsi = "http://www.w3.org/2001/XMLSchema-instance"
         xsi:schemaLocation = "http://maven.apache.org/POM/4.0.0 http://maven.apache.org/xsd/maven-4.0.0.xsd">
    <modelVersion>4.0.0</modelVersion>
    <groupId>com.bookcode</groupId>
    <artifactId>springcloud-helloworlds</artifactId>
    <version>0.0.1-SNAPSHOT</version>
    <packaging>jar</packaging>
    <name>springcloud-helloworlds</name>
    <description>Demo project for Spring Cloud</description>
    <parent>
        <groupId>org.springframework.boot</groupId>
        <artifactId>spring-boot-starter-parent</artifactId>
        <!-- 可修改 Spring Boot 版本信息 -->
        <version>2.1.1.RELEASE</version>
        <relativePath/> <!-- lookup parent from repository -->
    </parent>
    <properties>
        <project.build.sourceEncoding>UTF-8</project.build.sourceEncoding>
        <project.reporting.outputEncoding>UTF-8</project.reporting.outputEncoding>
        <java.version>1.8</java.version>
        <!-- 可修改 Spring Cloud 版本信息 -->
        <spring-cloud.version>Finchley.RELEASE</spring-cloud.version>
    </properties>
    <dependencies>
        <dependency>
            <groupId>org.springframework.cloud</groupId>
            <artifactId>spring-cloud-starter-gateway</artifactId>
        </dependency>
        <dependency>
            <groupId>org.springframework.boot</groupId>
            <artifactId>spring-boot-starter-test</artifactId>
            <scope>test</scope>
        </dependency>
```

```xml
        </dependencies>
    <dependencyManagement>
        <dependencies>
            <dependency>
                <groupId>org.springframework.cloud</groupId>
                <artifactId>spring-cloud-dependencies</artifactId>
                <version>${spring-cloud.version}</version>
                <type>pom</type>
                <scope>import</scope>
            </dependency>
        </dependencies>
    </dependencyManagement>
    <build>
        <plugins>
            <plugin>
                <groupId>org.springframework.boot</groupId>
                <artifactId>spring-boot-maven-plugin</artifactId>
            </plugin>
        </plugins>
    </build>
    <repositories>
        <repository>
            <id>spring-milestones</id>
            <name>Spring Milestones</name>
            <url>https://repo.spring.io/milestone</url>
            <snapshots>
                <enabled>false</enabled>
            </snapshots>
        </repository>
    </repositories>
</project>
```

至此，完成了项目的创建工作，在此基础上就可以进行 Spring Cloud 开发了。为了内容的简洁，在本书后面章节的示例和案例中将不再介绍项目的创建过程。如果不太清楚 Spring Cloud 项目创建过程，请先熟悉本节内容，或者直接参考源代码中每个项目中 pom.xml 文件内容。

1.4.2 创建项目的基本构成情况

IDEA 创建完项目之后，项目中目录和文件的构成情况如图 1-6 所示。项目中的目录、文件可以分为三大部分。其中，src/main/java 目录下包括主程序入口类 SpringcloudHelloworldsApplication，可以通过运行该类来运行程序，开发时需要在此目录下添加所需的接口、类等文件。src/main/resources 是资源目录，用于存放项目的资源。例如，配置文件 application.properties。src/test 是单元测试目录，自动生成的测试文件 SpringcloudHelloworldsApplicationTests 位于该目录下，用于测试程序。另外，pom.xml 文件是项目管理（特别是管理项目依赖）的重要文件。

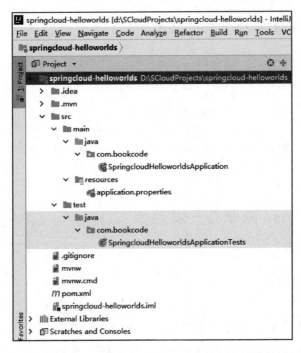

图1-6 IDEA 创建项目后项目的目录和文件的构成情况

1.4.3 基于 Spring Boot 的微服务实现

IDEA 创建完项目之后，可以开始实现微服务。Spring Cloud 基于 Spring Boot 来实现微服务。

在 com.bookcode 中创建 controller 子包，并在 com.bookcode.controller 中创建类 HelloController，修改类 HelloController 代码之后，类 HelloController 的代码如例1-2 所示。

【例1-2】 类 HelloController 修改后的代码示例。

```
package com.bookcode.controller;
import org.springframework.web.bind.annotation.GetMapping;
import org.springframework.web.bind.annotation.RestController;
@RestController
public class HelloController {
    @GetMapping("/hello")
```

```
    public String hello() {
        return "Hello, I am a Service.";
    }
}
```

运行程序,在浏览器中输入 localhost:8080/hello,结果如图 1-7 所示。

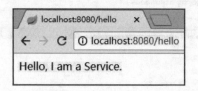

图 1-7   在浏览器中输入 localhost:8080/hello 后的结果

1.5 Spring Cloud 微服务开发起步

1.5.1 软件生命周期

软件工程是指导如何进行计算机软件开发和维护的一门工程学科。采用工程的概念、原理、技术和方法来开发与维护软件,把有效的管理技术和适用的技术方法结合起来,经济地开发出高质量的软件并有效地维护它,这就是软件工程。

按照软件工程的理论,软件生命周期由软件定义、软件开发和运行维护这 3 个时期组成,每个时期又进一步划分成若干个阶段。

软件定义时期的任务是确定软件项目开发必须完成的总目标;确定项目的可行性;导出实现项目开发目标应该采用的策略及系统必须完成的功能;估计完成项目需要的资源和成本,并且制订项目进度表。软件定义时期通常可进一步划分成问题定义、可行性研究和需求分析 3 个阶段。

软件开发时期的任务是设计和实现在前一个时期定义的软件,它通常由总体设计、详细设计、编码和单元测试、综合测试这 4 个阶段组成。

运行维护时期的主要任务是使软件持久地满足用户的需要。通常维护活动包括改正性维护、适应性维护、完善性维护和预防性维护。

本书主要探讨基于 Spring Cloud 的微服务开发,所以本书的实例和案例主要说明的是如何用 Spring Cloud 进行编码实现(简称为 Spring Cloud 开发)。

1.5.2 Spring Cloud 响应服务请求的处理顺序

在实现微服务后,就可以接收来自客户端(用户、消费者)的请求。收到客户端对服务的请求后,响应服务请求的一般处理顺序如下所述。

(1) 网关进行路由处理,即将请求转发给真正的服务提供者。

(2) 为了找到真正的服务提供者(即发现服务),需要服务提供者先注册服务。

(3) 发现服务,不一定能使用服务;用户要想能使用服务,须先要通过用户权限的验证和鉴定;于是,服务者在所提供的服务中需要增加认证和鉴权等安全功能。

(4) 通过验证之后,用户可以访问所需的服务;为了提高访问服务时的性能,需要进行负载均衡(可被看作是一种特殊的路由)和通过断路器实现容错机制(当一个服务发生故障、出现异常时断开该服务)。

(5) 服务提供者为了更好地管理配置信息,往往需要实现统一的配置管理机制。

(6) 服务提供者为了更好地管理服务,往往需要对服务进行监控、跟踪。

(7) 服务(请求者)与服务(响应者)之间往往需要进行消息处理。

(8) 服务的其他事项(如任务、数据流、多语言等)。

1.5.3 Spring Cloud 微服务开发的步骤

由于 Spring Cloud 开发是基于 Spring Boot 进行的,Spring Cloud 微服务响应服务请求的每项处理方案(即 1.5.2 节的一个步骤)的开发步骤基本相同。因此,Spring Cloud 微服务开发的步骤如下所述。

步骤 1:打开开发工具。

步骤 2:创建项目。

步骤 3:判断是否需要添加依赖;如果不需要,则跳过此步骤。

步骤 4:创建类、接口(按照实体类、数据访问接口和类、业务接口和类、控制器类等顺序)。

步骤 5:判断是否需要创建视图文件和 CSS 等文件;如果不需要,则跳过此步骤。

步骤 6:判断是否需要创建、修改配置文件;如果不需要,则跳过此步骤。

步骤 7:判断是否需要图片、语音、视频等文件;如果不需要,则跳过此步骤。

步骤8：判断是否需要下载辅助文件、包和安装工具（如数据库MySQL）；如果不需要，则跳过此步骤。注意，步骤3至步骤8之间的顺序可以互换。

步骤9：按照微服务架构的需要，根据情况再次执行步骤2至步骤8，直到不再需要新的开发循环为止。

步骤10：完成Spring Cloud开发之后，就可以运行程序了。由于微服务开发涉及的应用项目较多，在运行程序时，需要注意项目的运行次序。由于本书使用不同的端口来模拟多个应用。在运行程序时，需要注意项目所用的端口号和端口的启动顺序。

习题1

问答题

1. 请简述Spring的构成。
2. 请简述Spring Cloud的构成。
3. 请简述Spring Boot的特点。
4. 请简述与Spring Boot相比，Spring Cloud有哪些不同的特点。
5. 请简述Spring Cloud响应服务请求的处理顺序。
6. 请简述Spring Cloud微服务开发的步骤。

第 2 章

Spring Cloud 路由的应用

本章先介绍服务网关(也称为 API 网关)与 Spring Cloud 路由,再介绍 Spring Cloud Gateway、Spring Cloud Feign、Spring Cloud Ribbon、Spring Cloud Zuul 的应用。

2.1 服务网关与 Spring Cloud 路由简介

本节先介绍服务网关的作用,在此基础上,介绍 Spring Cloud 路由的相关技术。

2.1.1 服务网关的作用

微服务一般提供细粒度的 API(或称为功能、服务),而客户端通常需要和多个服务进行交互。服务网关是微服务架构中不可或缺的一部分。服务网关在统一向外提供系统服务的过程中,除了具备服务路由、均衡负载功能之外,还具备权限控制等功能。服务网关能隔离客户端和微服务,向客户端隐藏应用服务的划分和集成细节,并向每个客户端提供最优的服务。

随着业务需求的变化和时间的推移,服务网关背后的应用服务可能需要重新划分、实现和集成,这种调整和升级应该对客户端透明。这是软件中透明原则的体现。因此服务网关能够起到隐藏应用服务变化的作用。

服务网关也存在一些缺点。首先,实现、部署和管理网关需要耗费人力、物力、财力。其次,网关在客户端和服务之间多加了一层网络跳转,会给性能带来一定的损失(这一问题可以通过其他技术来弥补)。这些缺点与服务网关的优点相比显得微不足道。因此,在目前的应用中广泛使用服务网关。

以用于实现网关的 Spring Cloud Gateway 为例进行分析,其工作原理如图 2-1 所示。客户端向 Spring Cloud Gateway 发出请求,在 Gateway Handler Mapping 中找到与请求相匹配的路由,将其发送到 Gateway Web Handler;Handler 通过指定的过滤器链(多个过滤器)将请求发送给被网关代理的真正服务(即服务提供者),服务执行业务逻辑后返回结果。图 2-1 中每个过滤器内部的虚线表示过滤器可能会在发送代理请求之前(pre)或之后(post)执行业务逻辑。

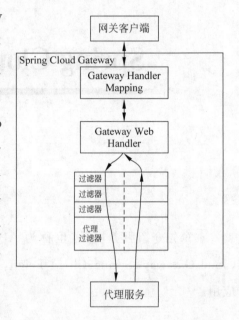

图 2-1 Spring Cloud Gateway 工作流程图

网关的主要作用是路由和过滤。并且网关具有安全、访问控制等功能。因此,在响应客户端请求时,首先是服务网关。在进行网关开发时,需要实现路由、过滤器等功能。

2.1.2 Spring Cloud 路由的相关技术

路由是确定服务最终的真正提供者的过程。Spring Cloud Gateway 旨在为微服务架构提供一种简单、有效、统一的 API 路由管理方式。因此,Spring Cloud 将服务网关的实现命名为 Spring Cloud Routing(路由)。此处的路由概念是广义上的路由。从这个角度来看,负载均衡也是一种路由方案;只不过负载均衡的出发点是服务分发而不是解决路由问题。因此,Spring Cloud 将实现负载均衡功能的 Ribbon 和 Feign(它集成了 Ribbon)也归并到 Spring Cloud Routing 中。

于是,广义地说,Spring Cloud 路由技术就包括了 Spring Cloud Gateway、Zuul、Ribbon 和 Feign 等技术。

2.2 Spring Cloud Gateway 路由的应用

视频讲解

Spring Cloud Gateway 作为 Spring Cloud 生态系统中的网关，目标是替代 Netflix Zuul。本节结合实例介绍 Spring Cloud Gateway 的路由开发。此处的路由概念是狭义的路由，是服务网关的一个组成部分。

2.2.1 创建项目并添加依赖

在 1.4.1 节介绍方法的基础上利用 IDEA 重新创建项目 springcloud-helloworlds，也可以在 1.4.1 节所创建项目的基础上直接进行开发。创建项目时，要添加 Spring Cloud Gateway 依赖。假如创建项目时没有添加 Spring Cloud Gateway 依赖，也可以在文件 pom.xml 的<dependencies>和</dependencies>之间直接添加 Spring Cloud Gateway 依赖，要添加的代码如例 2-1 所示。

【例 2-1】 在文件 pom.xml 的<dependencies>和</dependencies>之间直接添加依赖的代码示例。

```xml
<dependency>
    <groupId>org.springframework.cloud</groupId>
    <artifactId>spring-cloud-starter-gateway</artifactId>
</dependency>
```

2.2.2 创建类 HelloController

在包 com.bookcode 中创建 controller 子包，并在包 com.bookcode.controller 中创建类 HelloController，修改类 HelloController 代码后，类 HelloController 的代码如例 2-2 所示。一般在创建类之后都需要修改类代码，后面的章节将创建并修改类代码，简称为创建类。假如是在 1.4 节创建项目、实现服务的基础上直接进行开发，直接修改类 HelloController 的代码即可。

【例 2-2】 类 HelloController 的代码示例。

```java
package com.bookcode.controller;
import org.springframework.web.bind.annotation.GetMapping;
import org.springframework.web.bind.annotation.RestController;
@RestController                    //指定返回的默认结果为字符串
```

```
public class HelloController {
    @GetMapping("/hello")              //指定映射路径,使用 Get 方法
    public String hello() {
        return "Hello,我是 Spring Cloud Gateway.已经由 9000 端口跳转到新的端口.";
    }
}
```

2.2.3 修改配置文件 application.properties

修改目录 src/main/resources 下的配置文件 application.properties,修改后的代码如例 2-3 所示。

【例 2-3】 修改后的配置文件 application.properties 的代码示例。

```
#设置端口号
server.port = 9000
```

2.2.4 修改入口类

修改入口类,修改后的代码如例 2-4 所示。

【例 2-4】 修改后的入口类的代码示例。

```
package com.bookcode;
import org.springframework.boot.SpringApplication;
import org.springframework.boot.autoconfigure.SpringBootApplication;
import org.springframework.cloud.gateway.route.RouteLocator;
import org.springframework.cloud.gateway.route.builder.RouteLocatorBuilder;
import org.springframework.context.annotation.Bean;
@SpringBootApplication
public class SpringcloudHelloworldsApplication {
    //实现路由功能
    @Bean
    public RouteLocator customRouteLocator(RouteLocatorBuilder builder) {
        return builder.routes()
                .route(t -> t.path("/hello")
                        .and()
                        .uri("http://localhost:8080"))
                .build();
    }
    public static void main(String[] args) {
        SpringApplication.run(SpringcloudHelloworldsApplication.class, args);
    }
}
```

2.2.5 运行程序

运行程序后,控制台的输出结果如图 2-2 所示。控制台的输出由 3 部分构成:第一部分是路由谓词工厂加载信息(图 2-2 中第 1 行至第 13 行);第二部分是内嵌 Netty 服务器的启动信息(图 2-2 中第 14 行),请注意启动的端口是 9000;第三部分是应用启动时间(图 2-2 中第 15 行)。运行程序后,在浏览器中输入 localhost:9000/hello,结果如图 2-3 所示。

```
Loaded RoutePredicateFactory [After]
Loaded RoutePredicateFactory [Before]
Loaded RoutePredicateFactory [Between]
Loaded RoutePredicateFactory [Cookie]
Loaded RoutePredicateFactory [Header]
Loaded RoutePredicateFactory [Host]
Loaded RoutePredicateFactory [Method]
Loaded RoutePredicateFactory [Path]
Loaded RoutePredicateFactory [Query]
Loaded RoutePredicateFactory [ReadBodyPredicateFactory]
Loaded RoutePredicateFactory [RemoteAddr]
Loaded RoutePredicateFactory [Weight]
Loaded RoutePredicateFactory [CloudFoundryRouteService]
Netty started on port(s): 9000
Started SpringcloudHelloworldsApplication in 7.303 seconds (JVM running for 12.567)
```

图 2-2 控制台的输出结果

图 2-3 在浏览器中输入 localhost:9000/hello 后的结果

2.2.6 程序扩展

除了可以在入口类中设置路由信息,还可以在配置文件中设置路由信息。在目录 src/main/resources 下创建并修改配置文件 application.yml,修改后的代码如例 2-5 所示。一般在创建配置文件 application.yml 以后,都会向所创建的空白配置文件中添加代码(即修改配置文件),将此创建并修改配置文件的过程简称为创建配置文件。后面的章节中创建配置文件 application.yml 的示例代码均指修改后的代码。

【例 2-5】 创建配置文件 application.yml 的代码示例。

```yaml
spring:
  cloud:
    gateway:
      routes:
      - id: 163_route
        uri: http://www.163.com/
        predicates:
        - Path=/163
      - id: baidu_route
        uri: http://baidu.com:80/
        predicates:
        - Path=/baidu
```

运行程序后,在浏览器中输入 localhost:9000/hello,结果如图 2-3 所示。在浏览器中输入 localhost:9000/163,则成功跳转到网易首页(https://www.163.com/)。在浏览器中输入 localhost:9000/baidu,则跳转到百度首页(https://www.baidu.com/)。

2.3 Spring Cloud Gateway 过滤器的应用

视频讲解

本节介绍如何实现 Spring Cloud Gateway 过滤器的应用。

2.3.1 创建项目并添加依赖

用 IDEA 创建完项目 GatewayFilter 之后,确保在文件 pom.xml 的<dependencies>和</dependencies>之间添加了 Spring Cloud Gateway 依赖。

2.3.2 创建类 HelloController

在包 com.bookcode 中创建 controller 子包,并在包 com.bookcode.controller 中创建类 HelloController,修改代码后,类 HelloController 的代码如例 2-2 所示。后面的章节除非特别说明,创建类均包括创建类并修改类代码两个步骤,创建类的示例代码均指修改后的代码。

2.3.3 创建类 ElapsedFilter

在包 com.bookcode 中创建 filter 子包,并在包 com.bookcode.filter 中创建类

ElapsedFilter,代码如例 2-6 所示。

【例 2-6】 创建类 ElapsedFilter 的代码示例。

```java
package com.bookcode.filter;
import org.apache.commons.logging.Log;
import org.apache.commons.logging.LogFactory;
import org.springframework.cloud.gateway.filter.GatewayFilter;
import org.springframework.cloud.gateway.filter.GatewayFilterChain;
import org.springframework.core.Ordered;
import org.springframework.web.server.ServerWebExchange;
import reactor.core.publisher.Mono;
public class ElapsedFilter implements GatewayFilter, Ordered {
    private static final Log log = LogFactory.getLog(GatewayFilter.class);
    private static final String ELAPSED_TIME_BEGIN = "elapsedTimeBegin";
    @Override
    public Mono<Void> filter(ServerWebExchange exchange, GatewayFilterChain chain) {
        exchange.getAttributes().put(ELAPSED_TIME_BEGIN, System.currentTimeMillis());
        return chain.filter(exchange).then(
                Mono.fromRunnable(() -> {
                    Long startTime = exchange.getAttribute(ELAPSED_TIME_BEGIN);
                    if (startTime != null) {
                        log.info(exchange.getRequest().getURI().getRawPath().
toUpperCase() + ": " + (System.currentTimeMillis() - startTime) + "ms");
                    }
                })
        );
    }
    @Override
    public int getOrder() {
        return Ordered.LOWEST_PRECEDENCE;
    }
}
```

2.3.4 修改入口类

修改入口类,修改后的代码如例 2-7 所示。

【例 2-7】 修改后的入口类的代码示例。

```java
package com.bookcode;
import com.bookcode.filter.ElapsedFilter;
import org.springframework.boot.SpringApplication;
import org.springframework.boot.autoconfigure.SpringBootApplication;
import org.springframework.cloud.gateway.route.RouteLocator;
import org.springframework.cloud.gateway.route.builder.RouteLocatorBuilder;
import org.springframework.context.annotation.Bean;
@SpringBootApplication
```

```java
public class DemoApplication {
    @Bean
    public RouteLocator customerRouteLocator(RouteLocatorBuilder builder) {
        return builder.routes()
                .route(r -> r.path("/fluent/customer/**")
                        .filters(f -> f.stripPrefix(2)
                                .filter(new ElapsedFilter())
                                .addResponseHeader("X-Response-Default-Foo", "Default-Bar")
                                .setPath("/hello")
                        )
                        .uri("http://localhost:8080")
                        .order(0)
                )
                .build();
    }
    public static void main(String[] args) {
        SpringApplication.run(DemoApplication.class, args);
    }
}
```

2.3.5 运行程序

运行程序,在浏览器中输入 localhost:8080/fluent/customer/hello/zs,浏览器中的结果如图 2-4 所示。同时,控制台中的结果如图 2-5 所示。

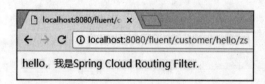

图 2-4 在浏览器中输入 localhost:8080/fluent/customer/hello/zs 后浏览器中的结果

图 2-5 在浏览器中输入 localhost:8080/fluent/customer/hello/zs 后控制台中的结果

2.4 Spring Cloud Feign 的应用

视频讲解

Feign 是一种模板化的 HTTP 客户端开发的工具,它的灵感来自于 Retrofit、JAX-RS(Java API for RESTful Web Services)和 WebSocket。在 Spring Cloud 中使用 Feign 可以实现用 HTTP 访问远程服务,就如同调用本地方法一样。

2.4.1 创建项目并添加依赖

用 IDEA 创建完项目 feignexample 之后,确保在文件 pom.xml 的 < dependencies > 和 </dependencies > 之间添加了 Web 和 Openfeign 依赖,代码如例 2-8 所示。

【例 2-8】 添加 Web 和 Feign 依赖的代码示例。

```xml
<dependency>
        <groupId>org.springframework.boot</groupId>
        <artifactId>spring-boot-starter-web</artifactId>
</dependency>
<dependency>
        <groupId>org.springframework.cloud</groupId>
        <artifactId>spring-cloud-starter-openfeign</artifactId>
</dependency>
```

2.4.2 创建接口 FeignService

在包 com.bookcode 中创建 service 子包,并在包 com.bookcode.service 中创建接口 FeignService,代码如例 2-9 所示。一般在创建接口之后都需要修改接口的代码,后面的章节将创建接口并修改代码的过程简称为创建接口。除非特别说明,创建接口的示例代码均指修改后的代码。

【例 2-9】 创建接口 FeignService 的代码示例。

```java
package com.bookcode.service;
import org.springframework.cloud.openfeign.FeignClient;
import org.springframework.web.bind.annotation.RequestMapping;
import org.springframework.web.bind.annotation.RequestMethod;
import org.springframework.web.bind.annotation.RequestParam;
@FeignClient(name = "github-client", url = "https://api.github.com")
public interface FeignService {
    @RequestMapping(value = "/search/repositories", method = RequestMethod.GET)
    String searchRepo(@RequestParam("q") String queryStr);
}
```

2.4.3 创建类 FeignController

在包 com.bookcode 中创建 controller 子包,并在包 com.bookcode.controller 中创建类 FeignController,代码如例 2-10 所示。

【例 2-10】 创建类 FeignController 的代码示例。

```java
package com.bookcode.controller;
import com.bookcode.service.FeignService;
import org.springframework.web.bind.annotation.GetMapping;
import org.springframework.web.bind.annotation.RequestParam;
import org.springframework.web.bind.annotation.RestController;
import javax.annotation.Resource;
@RestController
public class FeignController {
    @Resource
    private FeignService feignService;
    @GetMapping(value = "/search/github")
    public String searchGitHubRepoByStr(@RequestParam("str") String queryStr) {
        return feignService.searchRepo(queryStr);
    }
}
```

2.4.4 修改入口类

修改入口类，修改后的代码如例 2-11 所示。

【例 2-11】 修改后的入口类的代码示例。

```java
package com.bookcode;
import org.springframework.boot.SpringApplication;
import org.springframework.boot.autoconfigure.SpringBootApplication;
import org.springframework.cloud.openfeign.EnableFeignClients;
@EnableFeignClients
@SpringBootApplication
public class DemoApplication {
    public static void main(String[] args) {
        SpringApplication.run(DemoApplication.class, args);
    }
}
```

2.4.5 运行程序

运行程序后，在浏览器中输入 localhost:8080/search/github? str = spring-cloud，结果如图 2-6 所示。注意，由于网上内容会变化，读者运行后的结果可能会有差异。

第 2 章 Spring Cloud 路由的应用

图 2-6 在浏览器中输入 localhost:8080/search/github? str=spring-cloud 后的结果

2.5 Spring Cloud Ribbon 的应用

视频讲解

Spring Cloud Ribbon 是一个基于 HTTP 和 TCP 的客服端负载均衡工具，它是基于 Netflix Ribbon 实现的。它几乎可以用于每个微服务的基础设施中，因为负载均衡是提升系统可用性、缓解网络压力和扩容处理能力的重要手段之一。

2.5.1 创建项目并添加依赖

用 IDEA 创建完项目 ribbonexample 之后，确保在文件 pom.xml 的 < dependencies > 和</dependencies >之间添加了 Ribbon 和 Web 依赖，代码如例 2-12 所示。

【例 2-12】 添加 Ribbon 和 Web 依赖的代码示例。

```
< dependency >
         < groupId > org.springframework.cloud </groupId >
```

```
            <artifactId>spring-cloud-starter-netflix-ribbon</artifactId>
</dependency>
<dependency>
            <groupId>org.springframework.boot</groupId>
            <artifactId>spring-boot-starter-web</artifactId>
</dependency>
```

2.5.2 创建类 HelloController

在包 com.bookcode 中创建 controller 子包，并在包 com.bookcode.controller 中创建类 HelloController，代码如例 2-13 所示。

【例 2-13】 创建类 HelloController 的代码示例。

```java
package com.bookcode.controller;
import org.springframework.web.bind.annotation.GetMapping;
import org.springframework.web.bind.annotation.RestController;
import org.springframework.web.client.RestTemplate;
import javax.annotation.Resource;
@RestController
public class HelloController {
    @Resource
    private RestTemplate restTemplate;
    @GetMapping("/hi")
    public String Hello(){
        return this.restTemplate.getForObject("http://serviceprovider",String.class);
    }
}
```

2.5.3 修改配置文件 application.properties

修改配置文件 application.properties，修改后的代码如例 2-14 所示。

【例 2-14】 修改后的配置文件 application.properties 的代码示例。

```
#设置端口号
server.port = 8082
#设置要访问的服务
serviceprovider.ribbon.listOfServers: http://www.163.com/, http://baidu.com:80/
```

2.5.4 修改入口类

修改入口类，修改后的代码如例 2-15 所示。

【例 2-15】 修改后的入口类的代码示例。

```
package com.bookcode;
import org.springframework.boot.SpringApplication;
import org.springframework.boot.autoconfigure.SpringBootApplication;
import org.springframework.cloud.client.loadbalancer.LoadBalanced;
import org.springframework.context.annotation.Bean;
import org.springframework.web.client.RestTemplate;
@SpringBootApplication
public class DemoApplication {
    @Bean
    @LoadBalanced
    RestTemplate restTemplate() {
        return new RestTemplate();
    }
    public static void main(String[] args) {
        SpringApplication.run(DemoApplication.class, args);
    }
}
```

2.5.5 运行程序

运行程序后，在浏览器中输入 localhost:8082/hi 并不断进行刷新页面操作，结果会依次跳转到网易首页（https://www.163.com/）和百度首页（https://www.baidu.com/）。注意，每次运行结果可能会不同。

2.5.6 程序扩展

用 IDEA 创建完项目 serviceprovider 之后，确保在文件 pom.xml 的<dependencies>和</dependencies>之间添加了 Web 依赖，代码如例 2-16 所示。

【例 2-16】 添加 Web 依赖的代码示例。

```
<dependency>
        <groupId>org.springframework.boot</groupId>
        <artifactId>spring-boot-starter-web</artifactId>
</dependency>
```

在包 com.bookcode 中创建 controller 子包，并在包 com.bookcode.controller 中创建类 ProviderController，代码如例 2-17 所示。

【例 2-17】 创建类 ProviderController 的代码示例。

```
package com.bookcode.controller;
import org.springframework.beans.factory.annotation.Value;
import org.springframework.web.bind.annotation.RequestMapping;
import org.springframework.web.bind.annotation.RequestMethod;
import org.springframework.web.bind.annotation.RestController;
@RestController
public class ProviderController {
    @Value("${server.port}")
    String port;
    @RequestMapping(name = "/hello", method = RequestMethod.GET)
    public String index() {
        return "Hello SpringCloud PORT:" + port;
    }
}
```

修改配置文件 application.properties,修改后的代码如例 2-18 所示。

【例 2-18】 修改后的配置文件 application.properties 代码示例。

```
#设置端口号
server.port = 8080
spring.application.name = serviceprovider
```

运行程序后,在浏览器中输入 localhost:8080/hello,结果如图 2-7 所示。

保持项目 serviceprovider 中 pom.xml 文件、类 ProviderController 等内容不变,修改配置文件 application.properties,修改后的代码如例 2-19 所示。

【例 2-19】 修改后的配置文件 application.properties 代码示例。

```
#设置端口号
server.port = 8081
spring.application.name = serviceprovider
```

运行程序后,在浏览器中输入 localhost:8081/hello,结果如图 2-8 所示。

图 2-7 在浏览器中输入 localhost:8080/hello 后的结果

图 2-8 在浏览器中输入 localhost:8081/hello 后的结果

保持项目 ribbonexample 中 pom.xml 文件、类 HelloController、入口类不变,修改配置文件 application.properties,修改后的代码如例 2-20 所示。

【例 2-20】 修改后的配置文件 application.properties 代码示例。

```
#设置端口号
server.port = 8082
#设置要访问的服务
serviceprovider.ribbon.listOfServers:localhost:8080,localhost:8081
```

运行程序后,在浏览器中输入 localhost:8082/hi 并不断进行刷新页面操作,结果会依次如图 2-9(访问端口 8080)、图 2-10(访问端口 8081)所示。

图 2-9　在浏览器中输入 localhost:8082/hi 后的结果(访问端口 8080)

图 2-10　在浏览器中输入 localhost:8082/hi 后的结果(访问端口 8081)

2.6　Spring Cloud Zuul 的应用

视频讲解

Spring Cloud Zuul 担任了网关的角色,为微服务架构提供前门保护的作用,同时将权限控制等非业务逻辑内容迁移到服务路由层面,使得服务集群主体能够具备更高的可复用性和可测试性。

2.6.1　创建项目并添加依赖

用 IDEA 创建完项目 ZuulExample 之后,确保在文件 pom.xml 的< dependencies >和</dependencies >之间添加了 Zuul 依赖,代码如例 2-21 所示。

【例 2-21】 添加 Zuul 依赖的代码示例。

```
<dependency>
        <groupId>org.springframework.cloud</groupId>
        <artifactId>spring-cloud-starter-netflix-zuul</artifactId>
</dependency>
```

2.6.2　创建配置文件 application.yml

在目录 src/main/resources 下,创建配置文件 application.yml,代码如例 2-22 所示。

【例 2-22】 创建的配置文件 application.yml 代码示例。

```yaml
server:
  port: 9000
zuul:
  routes:
    163: #给路由任意起一个名字
      url: https://www.163.com
      path: /163/*
```

2.6.3 修改入口类

修改入口类,修改后的代码如例 2-23 所示。

【例 2-23】 修改后的入口类代码示例。

```java
package com.bookcode;
import org.springframework.boot.SpringApplication;
import org.springframework.boot.autoconfigure.SpringBootApplication;
import org.springframework.cloud.netflix.zuul.EnableZuulProxy;
@EnableZuulProxy
@SpringBootApplication
public class DemoApplication {
    public static void main(String[] args) {
        SpringApplication.run(DemoApplication.class, args);
    }
}
```

2.6.4 运行程序

运行程序后,在浏览器中输入 localhost:9000/163/,结果跳转到网易首页(https://www.163.com/)。

2.6.5 程序扩展

在包 com.bookcode 中创建 controller 子包,并在包 com.bookcode.controller 中创建类 HelloController,代码如例 2-24 所示。

【例 2-24】 创建类 HelloController 的代码示例。

```java
package com.bookcode.controller;
import org.springframework.web.bind.annotation.GetMapping;
```

```
import org.springframework.web.bind.annotation.RestController;
@RestController
public class HelloController {
    @GetMapping("/hello")
    public String hello() {
        return "Hello, I am Zuul.";
    }
}
```

修改配置文件 application.yml，修改后的代码如例 2-25 所示。

【例 2-25】 修改后的配置文件 application.yml 代码示例。

```
server:
  port: 9000
zuul:
  routes:
    163 : #给路由任意起一个名字
      url: https://www.163.com
      path: /163/*
    blog: #新增的路由
      url: http://localhost:8080
```

运行程序后，在浏览器中输入 localhost:9000/hello，结果如图 2-11 所示。

图 2-11　在浏览器中输入 localhost:9000/hello 后的结果

习题 2

一、问答题

1. 请简述网关的作用。
2. 请简述 Spring Cloud 路由的相关技术。

二、实验题

1. 请实现 Spring Cloud Gateway 的应用。
2. 请实现 Spring Cloud Feign 的应用。
3. 请实现 Spring Cloud Ribbon 的应用。
4. 请实现 Spring Cloud Zuul 的应用。

第 3 章

Spring Cloud 服务发现的应用

本章先简要介绍 Spring Cloud 的服务注册与发现,再说明如何实现 Spring Cloud Eureka、Spring Cloud Consul、Spring Cloud Zookeeper 的应用。

3.1 Spring Cloud 服务注册与发现的简介

客户的服务请求会通过 Spring Cloud 路由转发给真正的服务提供者。为了找到真正的服务提供者(即发现服务),需要服务提供者先注册服务。本节先介绍服务的注册和发现(简称服务发现);在此基础上,介绍 Spring Cloud 服务发现的相关解决方案。

3.1.1 服务的注册和发现

当一个系统存在大量微服务时,服务治理变得十分重要;因为各个服务需要通过服务治理来实现服务的自动化注册和发现。

在服务数量不多的情况下,可以通过配置中心的方法来实现服务治理。在使用此种方式情况下,当服务请求者需要调用某个服务时,可以基于配置中心中保存的目标服务地址完成调用。但是,这不是一种好的解决方案。为了实现微服务架构中的服务注册和发现,通常需要构建一个独立媒介来管理服务,这个媒介就是注册中心。

服务注册中心是服务提供者和服务消费者进行交互的媒介,具有服务注册和发现的作用。注册服务的目的是为了暴露服务访问的接口,此时注册中心的任务是存储服务的信息(如地址)。服务发现的目的是为了调用服务;如果服务调用超时或者失败,将需要采用集群容错机制。为了确保运行时的稳定性和可用性,还需要进行服务监控。

3.1.2　Spring Cloud 服务发现解决方案

Spring Cloud 服务发现解决方案包括 Spring Cloud Eureka、Spring Cloud Consul、Spring Cloud Zookeeper 和 Spring Cloud Foundry 等。

Spring Cloud Eureka 是对 Netflix 公司服务发现组件 Eureka 的封装,它包括 Eureka Server 和 Client。Eureka Server 提供 REST 服务,而 Eureka Client 是用 Java 编写的客户端,用于简化与 Eureka Server 的交互。

Spring Cloud Consul 是一个分布式解决方案,提供包括服务发现、配置和分段等功能。Consul 是以 HTTP 方式对外提供服务,提供了以服务治理为核心的多种功能。

Spring Cloud Zookeeper 提供一组简单的 API 使得开发人员可以实现通用的协作任务,包括选举主节点、管理组内成员关系、管理元数据等。服务可以被看作是一组连接到 ZooKeeper 服务器端的客户端,它们通过 ZooKeeper 客户端的 API 连接到 ZooKeeper 服务器端进行相关操作。

3.2　Spring Cloud Eureka 的应用

视频讲解

本节主要介绍 Spring Cloud Eureka 的开发,包括 Spring Cloud Eureka 服务器(即注册中心)和 Spring Cloud Eureka 客户端的开发。Spring Cloud Eureka 客户端又分为服务提供者(被调用者)和服务消费者(调用者)。Spring Cloud Eureka 客户端通过配置信息注册到注册中心。

3.2.1 Spring Cloud Eureka 注册中心的实现

用 IDEA 创建完项目 eureka-server 之后，确保在文件 pom.xml 的<dependencies>和</dependencies>之间添加了 Eureka Server 依赖，代码如例 3-1 所示。

【例 3-1】 添加 Eureka Server 依赖的代码示例。

```xml
<dependency>
    <groupId>org.springframework.cloud</groupId>
    <artifactId>spring-cloud-starter-netflix-eureka-server</artifactId>
</dependency>
```

在目录 src/main/resources 下，创建配置文件 application.yml，代码如例 3-2 所示。

【例 3-2】 创建的配置文件 application.yml 代码示例。

```yaml
server:
  port: 8761                        # 指定该 Eureka 实例的端口
eureka:
  instance:
    hostname: localhost
  client:
    registerWithEureka: false       # 表示不注册到 Eureka 注册中心,因为本应用是 Eureka 注册中心
    fetchRegistry: false
    serviceUrl:
      defaultZone: http://${eureka.instance.hostname}:${server.port}/eureka/
```

修改入口类，修改后的代码如例 3-3 所示。

【例 3-3】 修改后的入口类的代码示例。

```java
package com.bookcode;
import org.springframework.boot.SpringApplication;
import org.springframework.boot.autoconfigure.SpringBootApplication;
import org.springframework.cloud.netflix.eureka.server.EnableEurekaServer;
@SpringBootApplication
@EnableEurekaServer
public class DemoApplication {
    public static void main(String[] args) {
        SpringApplication.run(DemoApplication.class, args);
    }
}
```

3.2.2　Spring Cloud Eureka 服务提供者的实现

用 IDEA 创建完项目 eureka-client 之后，确保在文件 pom.xml 的<dependencies>和</dependencies>之间添加了驱动依赖，代码如例 3-4 所示。

【例 3-4】　添加 Eureka Client、Web、Data JPA、MySQL 数据库驱动依赖的代码示例。

```
<dependency>
        <groupId>org.springframework.cloud</groupId>
        <artifactId>spring-cloud-starter-netflix-eureka-client</artifactId>
</dependency>
<dependency>
        <groupId>org.springframework.boot</groupId>
        <artifactId>spring-boot-starter-web</artifactId>
</dependency>
<dependency>
        <groupId>org.springframework.boot</groupId>
        <artifactId>spring-boot-starter-data-jpa</artifactId>
</dependency>
<dependency>
        <groupId>mysql</groupId>
        <artifactId>mysql-connector-java</artifactId>
</dependency>
```

在包 com.bookcode 中创建 entity 子包，并在包 com.bookcode.entity 中创建类 Person，代码如例 3-5 所示。

【例 3-5】　创建类 Person 的代码示例。

```
package com.bookcode.entity;
import javax.persistence.*;
import java.math.BigDecimal;
@Entity
@Table(name = "person")
public class Person {
    @Id
    @GeneratedValue(strategy = GenerationType.IDENTITY)
    private Long id;
    @Column
    private String username;
    @Column
    private String name;
```

```java
    @Column
    private Integer age;
    @Column
    private BigDecimal balance;
    public Long getId() {
        return id;
    }
    public void setId(Long id) {
        this.id = id;
    }
    public String getUsername() {
        return username;
    }
    public void setUsername(String username) {
        this.username = username;
    }
    public String getName() {
        return name;
    }
    public void setName(String name) {
        this.name = name;
    }
    public Integer getAge() {
        return age;
    }
    public void setAge(Integer age) {
        this.age = age;
    }
    public BigDecimal getBalance() {
        return balance;
    }
    public void setBalance(BigDecimal balance) {
        this.balance = balance;
    }
    public String toString(){
        String ps = "Person : " + " id:" + getId() + ",username: " + getUsername() + ",name: " + getName() + ",age: " + getAge() + ",balance: " + getBalance() + "." ;
        return ps;
    }
}
```

在包 com.bookcode 中创建 dao 子包,并在包 com.bookcode.dao 中创建接口 PersonRepository,代码如例 3-6 所示。

【例 3-6】 创建接口 PersonRepository 的代码示例。

```java
package com.bookcode.dao;
import com.bookcode.entity.Person;
```

```
import org.springframework.data.jpa.repository.JpaRepository;
public interface PersonRepository extends JpaRepository<Person,Long> {
}
```

在包 com.bookcode 中创建 controller 子包,并在包 com.bookcode.controller 中创建类 PersonController,代码如例 3-7 所示。

【例 3-7】 创建类 PersonController 的代码示例。

```
package com.bookcode.controller;
import com.bookcode.dao.PersonRepository;
import com.bookcode.entity.Person;
import org.springframework.beans.factory.annotation.Autowired;
import org.springframework.beans.factory.annotation.Value;
import org.springframework.web.bind.annotation.PathVariable;
import org.springframework.web.bind.annotation.RequestMapping;
import org.springframework.web.bind.annotation.RestController;
import java.util.ArrayList;
import java.util.List;
@RestController
@RequestMapping("/person")
public class PersonController {
    @Autowired
    private PersonRepository personRepository;
    @Value("${server.port}")
    private String serverPort;
    @RequestMapping("/all")
    public List<String> getMemberAll() {
        List<String> listUser = new ArrayList<String>();
        listUser.add("赵高");
        listUser.add("李斯");
        listUser.add("胡亥");
        listUser.add("serverPort:" + serverPort);
        return listUser;
    }
    @RequestMapping("/{id}")
    public String findById(@PathVariable Long id){
        Person findOne = this.personRepository.getOne(id);
        return findOne.toString();
    }
    @RequestMapping("/getApi")
    public String getPersonApi(){
        return "这是 Eureka Client 的 Person 工程.";
    }
}
```

修改配置文件 application.properties,修改后的代码如例 3-8 所示。

【例3-8】 修改后的配置文件 application.properties 代码示例。

```
eureka.client.serviceUrl.defaultZone = http://localhost:8761/eureka/
server.port = 8762
spring.application.name = service-person
spring.datasource.url = jdbc:mysql://127.0.0.1:3306/springcloudtest
spring.datasource.username = root
spring.datasource.password = sa
spring.datasource.driver-class-name = com.mysql.jdbc.Driver
```

修改入口类,修改后的代码如例3-9所示。

【例3-9】 修改后的入口类的代码示例。

```java
package com.bookcode;
import org.springframework.boot.SpringApplication;
import org.springframework.boot.autoconfigure.SpringBootApplication;
@EnableEurekaClient        //从 Edgware 版 Spring Cloud 开始此行代码可省略
@SpringBootApplication
public class EurekaClientPersonApp {
    public static void main(String[] args) {
        SpringApplication.run(EurekaClientPersonApp.class, args);
    }
}
```

3.2.3 Spring Cloud Eureka 服务消费者的实现

用 IDEA 创建完项目 PersonServiceClient 之后,确保在文件 pom.xml 的 <dependencies> 和 </dependencies> 之间添加了 Eureka Client、Web 依赖,代码如例3-10所示。

【例3-10】 添加 Eureka Client、Web 依赖的代码示例。

```xml
<dependency>
        <groupId>org.springframework.cloud</groupId>
        <artifactId>spring-cloud-starter-netflix-eureka-client</artifactId>
</dependency>
<dependency>
        <groupId>org.springframework.boot</groupId>
        <artifactId>spring-boot-starter-web</artifactId>
</dependency>
```

在包 com.bookcode 中创建 service 子包,并在包 com.bookcode.service 中创建类 PersonService,代码如例3-11所示。

【例 3-11】 创建类 PersonService 的代码示例。

```java
package com.bookcode.service;
import java.util.List;
import org.springframework.stereotype.Service;
import org.springframework.web.client.RestTemplate;
import javax.annotation.Resource;
@Service
public class PersonService {
    @Resource
    private RestTemplate restTemplate;
    public List<String> getPersonAll() {
        return restTemplate.getForObject("http://service-person/person/all", List.class);
    }
}
```

在包 com.bookcode 中创建 controller 子包,并在包 com.bookcode.controller 中创建类 PersonServiceClientController,代码如例 3-12 所示。

【例 3-12】 创建类 PersonServiceClientController 的代码示例。

```java
package com.bookcode.controller;
import com.bookcode.service.PersonService;
import org.springframework.web.bind.annotation.RequestMapping;
import org.springframework.web.bind.annotation.RestController;
import javax.annotation.Resource;
import java.util.List;
@RestController
public class PersonServiceClientController {
    @Resource
    private PersonService personService;
    @RequestMapping("/getPesonAll")
    public List<String> getOrderUserAll() {
        return personService.getPersonAll();
    }
    @RequestMapping("/getOrderServiceApi")
    public String getOrderServiceApi(){
        return "这是服务 Person 消费者 Order 工程.";
    }
}
```

在目录 src/main/resources 下,创建配置文件 application.yml,代码如例 3-13 所示。

【例 3-13】 创建的配置文件 application.yml 代码示例。

```yaml
server:
  port: 8764
eureka:
  client:
    serviceUrl:
      defaultZone: http://localhost:8761/eureka/
spring:
  application:
    name: service-consumer
```

修改入口类,修改后的代码如例 3-14 所示。

【例 3-14】 修改后的入口类的代码示例。

```java
package com.bookcode;
import org.springframework.boot.SpringApplication;
import org.springframework.boot.autoconfigure.SpringBootApplication;
import org.springframework.cloud.client.loadbalancer.LoadBalanced;
import org.springframework.context.annotation.Bean;
import org.springframework.web.client.RestTemplate;
@SpringBootApplication
public class DemoApplication {
    @Bean
    @LoadBalanced       //实现负载负衡
    RestTemplate restTemplate() {
        return new RestTemplate();
    }
    public static void main(String[] args) {
        SpringApplication.run(DemoApplication.class, args);
    }
}
```

3.2.4 运行程序

运行注册中心程序 eureka-server,它的端口为 8761。运行服务提供者程序 eureka-client,端口为 8762。保持项目其他代码不变的情况下修改项目 eureka-client 的配置文件 application.properties,修改后的代码如例 3-15 所示。然后,新运行一个服务提供者程序 eureka-client,端口为 8763。运行服务消费者程序 PersonServiceClient,它的端口为 8764。

【例 3-15】 修改后的配置文件 application.properties 代码示例。

```
eureka.client.serviceUrl.defaultZone = http://localhost:8761/eureka/
server.port = 8763
```

```
spring.application.name = service-person
spring.datasource.url = jdbc:mysql://127.0.0.1:3306/springcloudtest
spring.datasource.username = root
spring.datasource.password = sa
spring.datasource.driver-class-name = com.mysql.jdbc.Driver
```

运行完程序后,在浏览器中输入 localhost:8761,结果如图 3-1 所示。结果显示了所有注册到 Eureka 服务器的服务(注意,SERVICE-PERSON 服务的端口号包括 8762 和 8763)。在浏览器中输入 localhost:8762/person/3,结果如图 3-2 所示。在浏览器中输入 localhost:8763/person/all,结果如图 3-3 所示。在浏览器中输入 localhost:8763/person/getApi,结果如图 3-4 所示。在浏览器中输入 localhost:8764/getPesonAll 并不断进行页面刷新操作,结果依次如图 3-5 和图 3-6 所示。在浏览器中输入 localhost:8764/getOrderServiceApi,结果如图 3-7 所示。

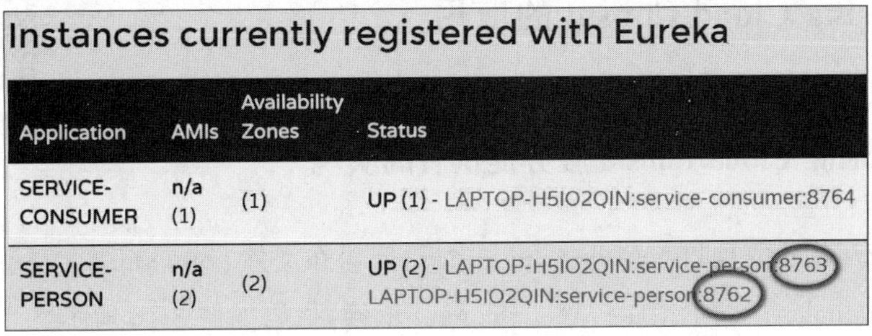

图 3-1 在浏览器中输入 localhost:8761 后的结果

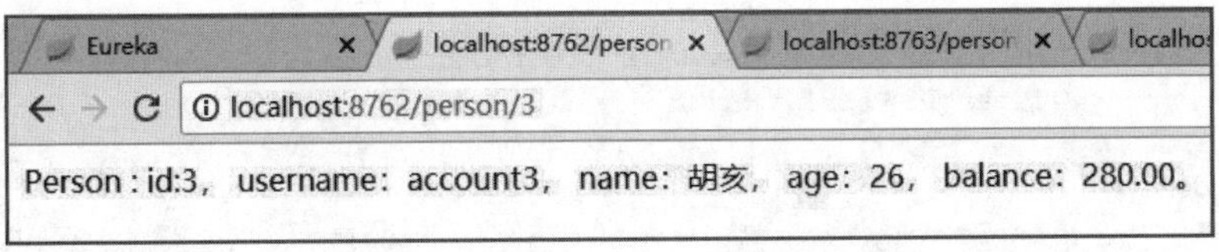

图 3-2 在浏览器中输入 localhost:8762/person/3 后的结果

图 3-3 在浏览器中输入 localhost:8763/
person/all 后的结果

图 3-4 在浏览器中输入 localhost:8763/
person/getApi 后的结果

图 3-5 在浏览器中输入 localhost:8764/
getPesonAll 后的结果(端口 8762)

图 3-6 在浏览器中输入 localhost:8764/
getPesonAll 后的结果(端口 8763)

图 3-7 在浏览器中输入 localhost:8764/getOrderServiceApi 后的结果

3.3 Spring Cloud Consul 的应用

视频讲解

3.3.1 Spring Cloud Consul 服务提供者的实现

用 IDEA 创建完项目 consulprovider 之后,确保在文件 pom.xml 的<dependencies>和</dependencies>之间添加了 Web、Consul 依赖,代码如例 3-16 所示。

【例 3-16】 添加 Web、Consul 依赖的代码示例。

```
<dependency>
        <groupId>org.springframework.boot</groupId>
        <artifactId>spring-boot-starter-web</artifactId>
</dependency>
<dependency>
        <groupId>org.springframework.cloud</groupId>
        <artifactId>spring-cloud-starter-consul-discovery</artifactId>
</dependency>
```

在包 com.bookcode 中创建 service 子包,并在包 com.bookcode.service 中创建接口 IHello,代码如例 3-17 所示。

【例 3-17】 创建接口 IHello 的代码示例。

```
package com.bookcode.service;
public interface IHello {
    public String sayHello(String name);
}
```

在包 com.bookcode.service 中创建 impl 子包，并在包 com.bookcode.service.impl 中创建类 HelloImpl，代码如例 3-18 所示。

【例 3-18】 创建类 HelloImpl 的代码示例。

```
package com.bookcode.service.impl;
import com.bookcode.service.IHello;
import org.springframework.stereotype.Service;
@Service
public class HelloImpl implements IHello {
    @Override
    public String sayHello(String name) {
        return "Hello, I am " + name + " .";
    }
}
```

在包 com.bookcode 中创建 controller 子包，并在包 com.bookcode.controller 中创建类 HelloController，代码如例 3-19 所示。

【例 3-19】 创建类 HelloController 的代码示例。

```
package com.bookcode.controller;
import com.bookcode.service.impl.HelloImpl;
import org.springframework.beans.factory.annotation.Autowired;
import org.springframework.web.bind.annotation.RequestMapping;
import org.springframework.web.bind.annotation.RestController;
@RestController
@RequestMapping("/hello")
public class HelloController {
    @Autowired
    private HelloImpl hello;
    @RequestMapping("/say")
    public String sayHello(String name){
        return hello.sayHello(name);
    }
}
```

在包 com.bookcode.controller 中创建类 HealthController，代码如例 3-20 所示。

【例 3-20】 创建类 HealthController 的代码示例。

```
package com.bookcode.controller;
import org.springframework.web.bind.annotation.RequestMapping;
import org.springframework.web.bind.annotation.RestController;
@RestController
public class HealthController {
    @RequestMapping("/health")
```

```
    public String health(){
        return "health";
    }
}
```

在目录 src/main/resources 下,创建配置文件 application.yml,代码如例 3-21 所示。

【例 3-21】 创建的配置文件 application.yml 代码示例。

```
spring:
  cloud:
    consul:
      host: localhost
      port: 8500
      discovery:
        healthCheckPath: /health
        healthCheckInterval: 15s
        instance-id: consul2
        enabled: true
      enabled: true
  application:
    name: consulprovider
server:
  port: 8082
```

修改入口类,修改后的代码如例 3-22 所示。

【例 3-22】 修改后的入口类的代码示例。

```
package com.bookcode;
import org.springframework.boot.SpringApplication;
import org.springframework.boot.autoconfigure.SpringBootApplication;
import org.springframework.cloud.client.discovery.EnableDiscoveryClient;
@EnableDiscoveryClient
@SpringBootApplication
public class DemoApplication {
    public static void main(String[] args) {
        SpringApplication.run(DemoApplication.class, args);
    }
}
```

3.3.2 Spring Cloud Consul 服务消费者的实现

用 IDEA 创建完项目 consulconsumer 之后,确保在文件 pom.xml 的<dependencies>和</dependencies>之间添加了 Web、Actuator、Openfeign、Consul 依赖,代码如例 3-23

所示。

【例 3-23】 添加 Web、Actuator、Openfeign、Consul 依赖的代码示例。

```xml
<dependency>
    <groupId>org.springframework.boot</groupId>
    <artifactId>spring-boot-starter-web</artifactId>
</dependency>
<dependency>
    <groupId>org.springframework.boot</groupId>
    <artifactId>spring-boot-starter-actuator</artifactId>
</dependency>
<dependency>
    <groupId>org.springframework.cloud</groupId>
    <artifactId>spring-cloud-starter-openfeign</artifactId>
</dependency>
<dependency>
    <groupId>org.springframework.cloud</groupId>
    <artifactId>spring-cloud-starter-consul-discovery</artifactId>
</dependency>
```

在包 com.bookcode 中创建 service 子包，并在包 com.bookcode.service 中创建接口 Greeting，代码如例 3-24 所示。

【例 3-24】 创建接口 Greeting 的代码示例。

```java
package com.bookcode.service;
import org.springframework.cloud.openfeign.FeignClient;
import org.springframework.web.bind.annotation.RequestMapping;
import org.springframework.web.bind.annotation.RequestMethod;
import org.springframework.web.bind.annotation.RequestParam;
@FeignClient("consulprovider")
public interface Greeting {
    @RequestMapping(value = "/hello/say", method = RequestMethod.GET)
    String sayHello(@RequestParam("name") String name);
}
```

在包 com.bookcode 中创建 controller 子包，并在包 com.bookcode.controller 中创建类 TestController，代码如例 3-25 所示。

【例 3-25】 创建类 TestController 的代码示例。

```java
package com.bookcode.controller;
import com.bookcode.service.Greeting;
import org.springframework.web.bind.annotation.RequestMapping;
import org.springframework.web.bind.annotation.RestController;
import javax.annotation.Resource;
```

```
@RestController
@RequestMapping("/test")
public class TestController {
    @Resource
    private Greeting greeting;
    @RequestMapping("/hi")
    public String testHi(String name){
        return greeting.sayHello(name);
    }
}
```

在包 com.bookcode.controller 中创建类 HealthController,代码如例 3-20 所示。

在目录 src/main/resources 下,创建配置文件 application.yml,代码如例 3-26 所示。

【例 3-26】 创建的配置文件 application.yml 代码示例。

```
spring:
  cloud:
    consul:
      host: localhost
      port: 8500
      discovery:
        healthCheckPath: /health
        healthCheckInterval: 15s
        instance-id: consul1
        enabled: true
      enabled: true
  application:
    name: consulconsumer
server:
  port: 8081
```

修改入口类,修改后的代码如例 3-27 所示。

【例 3-27】 修改后的入口类的代码示例。

```
package com.bookcode;
import org.springframework.boot.SpringApplication;
import org.springframework.boot.autoconfigure.SpringBootApplication;
import org.springframework.cloud.client.discovery.EnableDiscoveryClient;
import org.springframework.cloud.openfeign.EnableFeignClients;
@EnableDiscoveryClient
@EnableFeignClients
@SpringBootApplication
```

```
public class DemoApplication {
    public static void main(String[] args) {
        SpringApplication.run(DemoApplication.class, args);
    }
}
```

3.3.3 运行程序

安装 Consul，打开 Windows 命令处理程序 CMD，执行如例 3-28 所示的命令，启动 Consul 服务。

【例 3-28】 启动 Consul 服务的命令示例。

```
consul agent –dev
```

运行服务提供者程序 consulprovider，它的端口为 8082。运行服务消费者程序 consulconsumer，它的端口为 8081。

程序运行完成后，在浏览器中输入 localhost:8082/health，结果如图 3-8 所示。在浏览器中输入 localhost:8082/hello/say？name＝Consul，结果如图 3-9 所示。在浏览器中输入 localhost:8081/test/hi？name＝Consul，结果如图 3-10 所示。在浏览器中输入 localhost:8081/health，结果如图 3-11 所示。在浏览器中输入 localhost:8500/ui/dc1/services，结果如图 3-12 所示，显示所有注册和被发现了的服务。

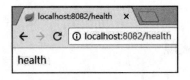

图 3-8　在浏览器中输入 localhost:8082/health 后的结果

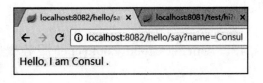

图 3-9　在浏览器中输入 localhost:8082/hello/say?name＝Consul 后的结果

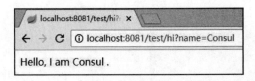

图 3-10　在浏览器中输入 localhost:8081/test/hi?name＝Consul 后的结果

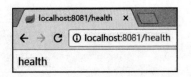

图 3-11　在浏览器中输入 localhost:8081/health 后的结果

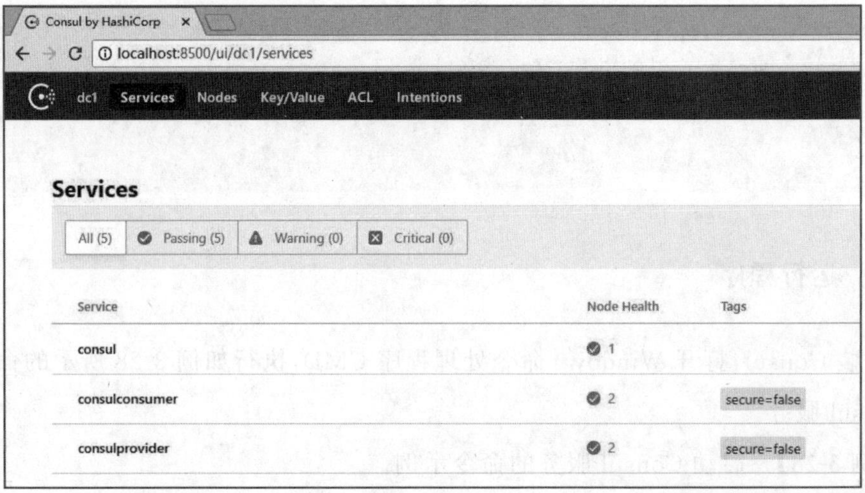

图 3-12　在浏览器中输入 localhost:8500/ui/dc1/services 后的结果

3.4　Spring Cloud Zookeeper 的应用

视频讲解

ZooKeeper 安装方式有 3 种，单机模式、集群模式和伪集群模式。在单机模式下，ZooKeeper 只运行在一台服务器上，适合测试环境。在集群模式下，ZooKeeper 运行于一个集群上，适合于生产环境。在伪集群模式下，在一台物理机上运行多个 ZooKeeper 实例。本节以单机模式为例介绍 Spring Cloud Zookeeper 的应用。

3.4.1　Spring Cloud Zookeeper 服务提供者的实现

用 IDEA 创建完项目 zookeeperprovider 之后，确保在文件 pom.xml 的<dependencies>和</dependencies>之间添加了 Web、Zookeeper 依赖，代码如例 3-29 所示。

【例 3-29】　添加 Web、Zookeeper 依赖的代码示例。

```xml
<dependency>
    <groupId>org.springframework.boot</groupId>
    <artifactId>spring-boot-starter-web</artifactId>
</dependency>
<dependency>
    <groupId>org.springframework.cloud</groupId>
    <artifactId>spring-cloud-starter-zookeeper-discovery</artifactId>
</dependency>
```

在包 com.bookcode 中创建 controller 子包,并在包 com.bookcode.controller 中创建类 HelloController,代码如例 3-30 所示。

【例 3-30】 创建类 HelloController 的代码示例。

```
package com.bookcode.controller;
import org.springframework.web.bind.annotation.RequestMapping;
import org.springframework.web.bind.annotation.RestController;
@RestController
public class HelloController {
    @RequestMapping("/hi")
    public String home() {
        return "Hello, I am ZooKeeper.";
    }
}
```

修改配置文件 application.properties,修改后的代码如例 3-31 所示。

【例 3-31】 修改后的配置文件 application.properties 的代码示例。

```
server.port = 8822
spring.application.name = spring-cloud-zookeeper-provider
spring.cloud.zookeeper.connectString = localhost:2181
```

修改入口类,修改后的代码如例 3-32 所示。

【例 3-32】 修改后的入口类的代码示例。

```
package com.bookcode;
import org.springframework.boot.SpringApplication;
import org.springframework.boot.autoconfigure.SpringBootApplication;
import org.springframework.cloud.client.discovery.EnableDiscoveryClient;
@EnableDiscoveryClient
@SpringBootApplication
public class DemoApplication {
    public static void main(String[] args) {
        SpringApplication.run(DemoApplication.class, args);
    }
}
```

3.4.2 Spring Cloud Zookeeper 服务消费者的实现

用 IDEA 创建完项目 zookeeperconsumer 之后,确保在文件 pom.xml 的 <dependencies>和</dependencies>之间添加了 Web、Zookeeper 依赖,代码如

例 3-33 所示。

【例 3-33】 添加 Web、Zookeeper 依赖的代码示例。

```xml
<dependency>
        <groupId>org.springframework.boot</groupId>
        <artifactId>spring-boot-starter-web</artifactId>
</dependency>
<dependency>
        <groupId>org.springframework.cloud</groupId>
        <artifactId>spring-cloud-starter-zookeeper-discovery</artifactId>
</dependency>
```

在包 com.bookcode 中创建 controller 子包,并在包 com.bookcode.controller 中创建类 TestController,代码如例 3-34 所示。

【例 3-34】 创建类 TestController 的代码示例。

```java
package com.bookcode.controller;
import org.springframework.beans.factory.annotation.Autowired;
import org.springframework.web.bind.annotation.RequestMapping;
import org.springframework.web.bind.annotation.RequestMethod;
import org.springframework.web.bind.annotation.RestController;
import org.springframework.web.client.RestTemplate;
@RestController
public class TestController {
    @Autowired
    RestTemplate restTemplate;
    @RequestMapping(value = "/hello", method = RequestMethod.GET)
    public String add() {
    return restTemplate.getForEntity("http://spring-cloud-zookeeper-provider/hi", String.class).getBody();
    }
}
```

修改配置文件 application.properties,修改后的代码如例 3-35 所示。

【例 3-35】 修改后的配置文件 application.properties 的代码示例。

```
server.port = 8833
spring.application.name = spring-cloud-zookeeper-client
spring.cloud.zookeeper.connectString = localhost:2181
spring.cloud.zookeeper.discovery.instanceHost = localhost
spring.cloud.zookeeper.discovery.instancePort = ${server.port}
```

修改入口类,修改后的代码如例 3-36 所示。

【例 3-36】 修改后的入口类的代码示例。

```
package com.bookcode;
import org.springframework.boot.SpringApplication;
import org.springframework.boot.autoconfigure.SpringBootApplication;
import org.springframework.cloud.client.discovery.EnableDiscoveryClient;
import org.springframework.cloud.client.loadbalancer.LoadBalanced;
import org.springframework.context.annotation.Bean;
import org.springframework.web.client.RestTemplate;
@EnableDiscoveryClient
@SpringBootApplication
public class DemoApplication {
    @Bean
    @LoadBalanced
    public RestTemplate restTemplate() {
        return new RestTemplate();
    }
    public static void main(String[] args) {
        SpringApplication.run(DemoApplication.class, args);
    }
}
```

3.4.3 运行程序

安装 ZooKeeper，打开 Windows 命令处理程序 CMD，执行如例 3-37 所示命令，启动 ZooKeeper 服务。

【例 3-37】 启动 ZooKeeper 服务的命令示例。

```
zkServer
```

运行服务提供者程序 zookeeperprovider，它的端口为 8822。运行服务消费者程序 zookeeperconsumer，端口为 8833。保持程序 zookeeperconsumer 其他代码不变的情况下修改配置文件 application.properties，修改后的代码如例 3-38 所示。再运行一个新的服务提供者程序 zookeeperconsumer，端口为 8844。

【例 3-38】 修改后的配置文件 application.properties 代码示例。

```
server.port = 8844
spring.application.name = spring-cloud-zookeeper-client
spring.cloud.zookeeper.connectString = localhost:2181
spring.cloud.zookeeper.discovery.instanceHost = localhost
spring.cloud.zookeeper.discovery.instancePort = ${server.port}
```

运行完程序后,在浏览器中输入 localhost：8822/hi,结果如图 3-13 所示。在浏览器中输入 localhost：8833/hello,结果如图 3-14 所示。在浏览器中输入 localhost：8844/hello,结果如图 3-15 所示。

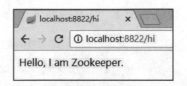

图 3-13　在浏览器中输入 localhost：8822/hi 后的结果

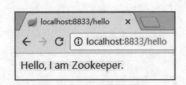

图 3-14　在浏览器中输入 localhost：8833/hello 后的结果

图 3-15　在浏览器中输入 localhost：8844/hello 后的结果

打开 Windows 命令处理程序 CMD,执行如例 3-39 所示命令,启动 ZooKeeper 客户端。

【例 3-39】启动 ZooKeeper 客户端的命令示例。

```
zkCli
```

注意,在后面章节中若没有明确地说明执行命令的环境,则均指在 Windows 命令行处理程序 CMD 中执行命令。

执行 ZooKeeper 命令查看服务注册情况,命令如例 3-40 所示,结果如图 3-16 所示,可以看到有一个服务提供者、两个服务消费者注册到 ZooKeeper。

【例 3-40】查看 ZooKeeper 服务注册的命令示例。

```
ls/
ls/services
ls/services/spring-cloud-zookeeper-client
ls/services/spring-cloud-zookeeper-provider
```

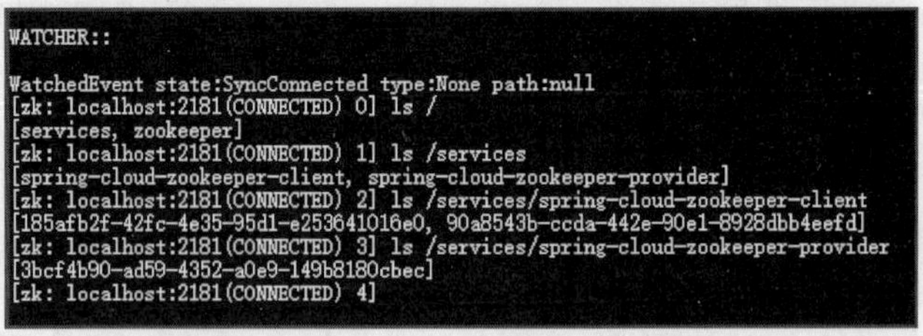

图 3-16　执行 ZooKeeper 命令查看服务注册情况的结果

习题 3

一、问答题

请简述服务的注册和发现的关系。

二、实验题

1. 请实现 Spring Cloud Eureka 的应用。

2. 请实现 Spring Cloud Consul 的应用。

3. 请实现 Spring Cloud Zookeeper 的应用。

第 4 章

Spring Cloud 认证与鉴权的应用

通俗地讲,认证就是验证用户是谁,鉴权是确定用户能做什么事。本章先介绍 Spring Cloud Security、Spring Cloud OAuth2(简称 OAuth2)、JWT(JSON Web Token)的简单应用;再介绍 Spring Cloud Gateway(简称 Gateway)、JWT、Actuator 的综合应用;最后,介绍 Eureka、Zuul、OAuth 和 JWT 的综合应用。

4.1　Spring Cloud Security 的简单应用

视频讲解

Spring Cloud Security 提供了一组原语,用于构建安全的应用程序和服务,而且操作简便。

4.1.1　创建项目并添加依赖

用 IDEA 创建完项目 securityexample 之后,确保在文件 pom.xml 的< dependencies >和</dependencies >之间添加了 Web 和 Security 依赖,代码如例 4-1 所示。

【例 4-1】　添加 Web 和 Security 依赖的代码示例。

```
< dependency >
```

```
            <groupId>org.springframework.boot</groupId>
            <artifactId>spring-boot-starter-web</artifactId>
</dependency>
<dependency>
            <groupId>org.springframework.cloud</groupId>
            <artifactId>spring-cloud-starter-security</artifactId>
</dependency>
```

4.1.2 创建类 HelloController

在包 com.bookcode 中创建 controller 子包,并在包 com.bookcode.controller 中创建类 HelloController,代码如例 4-2 所示。

【例 4-2】 创建类 HelloController 的代码示例。

```
package com.bookcode.controller;
import org.springframework.web.bind.annotation.GetMapping;
import org.springframework.web.bind.annotation.RestController;
@RestController
public class HelloController {
    @GetMapping("/hello")
    public String hello() {
        return "Hello, I am Spring Cloud Security.";
    }
}
```

4.1.3 创建配置文件 application.yml

在目录 src/main/resources 下,创建配置文件 application.yml,代码如例 4-3 所示。

【例 4-3】 创建的配置文件 application.yml 代码示例。

```
server:
  port: 8761
spring:
  security:
    basic:
      enabled: true        #开启基于 HTTP basic 的认证
    user:
      name: zs             #配置登录的账号
      password: zs         #配置登录的密码
```

4.1.4 运行程序

运行程序后,在浏览器中输入 localhost:8761/hello,页面跳转到 localhost:8761/

login,结果如图 4-1 所示。输入正确的 User 和 Password 后,页面跳转到 localhost:8761/hello,结果如图 4-2 所示。

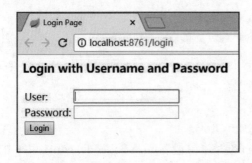

图 4-1 在浏览器中输入 localhost:8761/hello 后跳转到 localhost:8761/login 的结果

图 4-2 输入正确的 User 和 Password 后页面跳转到 localhost:8761/hello 的结果

4.1.5 程序扩展

在文件 pom.xml 的<dependencies>和</dependencies>之间添加 Eureka Server 依赖,代码如例 4-4 所示。

【例 4-4】 添加 Eureka Server 依赖的代码示例。

```
<dependency>
        <groupId>org.springframework.cloud</groupId>
        <artifactId>spring-cloud-starter-netflix-eureka-server</artifactId>
</dependency>
```

修改配置文件 application.yml,修改后的代码如例 4-5 所示。

【例 4-5】 修改后的配置文件 application.yml 代码示例。

```
server:
  port: 8761
spring:
  security:
    basic:
      enabled: true              #开启基于 HTTP basic 的认证
    user:
      name: zs                   #配置登录的账号
      password: zs               #配置登录的密码
eureka:
  client:
    registerWithEureka: false
```

```
    fetchRegistry: false
    serviceUrl:
      defaultZone: http://localhost:8761/eureka/
```

修改入口类，修改后的代码如例 4-6 所示。

【例 4-6】 修改后的入口类的代码示例。

```
package com.bookcode;
import org.springframework.boot.SpringApplication;
import org.springframework.boot.autoconfigure.SpringBootApplication;
import org.springframework.cloud.netflix.eureka.server.EnableEurekaServer;
@EnableEurekaServer
@SpringBootApplication
public class DemoApplication {
    public static void main(String[] args) {
        SpringApplication.run(DemoApplication.class, args);
    }
}
```

运行程序后，在浏览器中输入 localhost:8761，页面跳转到 localhost:8761/login，结果如图 4-1 所示。输入正确的 User 和 Password 后，结果如图 4-3 所示。然后，在浏览器中输入 localhost:8761/hello，结果如图 4-2 所示。

图 4-3 输入正确的 User 和 Password 后的结果

4.2　Spring Cloud OAuth 2 的简单应用

视频讲解

协议 OAuth 在客户端与服务提供者之间,设置了一个授权层 (authorization layer)。客户端不能直接登录服务提供者,只能登录授权层,以此将服务提供者与客户端区分开来。客户端登录时使用令牌(Token),指定授权层令牌的权限范围和有效期。客户端登录授权层以后,服务提供者根据令牌的权限范围和有效期,向客户端开放用户存储的资料。本节将介绍 Spring Cloud OAuth2 的简单应用;在应用前先需要在 GitHub 上新建一个应用,获取客户端的 Client ID 和 Client Secret 并配置好回调地址。

4.2.1　创建项目并添加依赖

用 IDEA 创建完项目 oauthexample 之后,确保在文件 pom.xml 的<dependencies>和</dependencies>之间添加了 OAuth2、Security 和 Web 依赖,代码如例 4-7 所示。

【例 4-7】　添加 OAuth2、Security 和 Web 依赖的代码示例。

```
<dependency>
    <groupId>org.springframework.cloud</groupId>
    <artifactId>spring-cloud-starter-oauth2</artifactId>
</dependency>
<dependency>
    <groupId>org.springframework.cloud</groupId>
    <artifactId>spring-cloud-starter-security</artifactId>
</dependency>
<dependency>
    <groupId>org.springframework.boot</groupId>
    <artifactId>spring-boot-starter-web</artifactId>
</dependency>
```

4.2.2　创建类 HelloController

在包 com.bookcode 中创建 controller 子包,并在包 com.bookcode.controller 中创建类 HelloController,代码如例 4-8 所示。

【例 4-8】　创建类 HelloController 的代码示例。

```
package com.bookcode.controller;
import org.springframework.web.bind.annotation.RequestMapping;
import org.springframework.web.bind.annotation.RestController;
@RestController
public class HelloController {
    @RequestMapping("/hi")
    public String index() {
        return "Hello, I am OAuth2 in Controller.";
    }
}
```

4.2.3 创建文件 index.html

在目录 src/main/resources/static 下，创建文件 index.html，代码如例 4-9 所示。

【例 4-9】 创建的文件 index.html 代码示例。

```
<h1>Hello, I am OAuth2 in HTML File!</h1>
```

4.2.4 创建配置文件 application.yml

在目录 src/main/resources 下，创建配置文件 application.yml，代码如例 4-10 所示。

【例 4-10】 创建的配置文件 application.yml 代码示例。

```
server:
  port: 7073
security:
  user:
    password: user                    #直接登录时的密码
  ignored: /
  sessions: never                     #session策略
  oauth2:
    sso:
      loginPath: /login               #登录路径
    client:
      clientId: 1db80261fb248dd7f669  #设计时需要改成用户自己的clientId
      clientSecret: 01b0f85f3cdb70d87473053d0f09fe1e133d719c
#设计时需要改成用户的clientSecret
      accessTokenUri: https://github.com/login/oauth/access_token
      userAuthorizationUri: https://github.com/login/oauth/authorize
    resource:
      userInfoUri: https://api.github.com/user
      preferTokenInfo: false
```

4.2.5 修改入口类

修改入口类，修改后的代码如例 4-11 所示。

【例 4-11】 修改后的入口类的代码示例。

```
package com.bookcode;
import org.springframework.boot.SpringApplication;
import org.springframework.boot.autoconfigure.SpringBootApplication;
import org.springframework.boot.autoconfigure.security.oauth2.client.EnableOAuth2Sso;
@EnableOAuth2Sso
@SpringBootApplication
public class DemoApplication {
    public static void main(String[] args) {
        SpringApplication.run(DemoApplication.class, args);
    }
}
```

4.2.6 运行程序

运行程序后，在浏览器中输入 127.0.0.1:7073，会跳转到 GitHub 的授权页面，结果如图 4-4 所示。登录并得到授权之后，跳转到 127.0.0.1:7073，结果如图 4-5 所示。在浏览器中输入 127.0.0.1:7073/hi，结果如图 4-6 所示。

图 4-4 输入 127.0.0.1:7073 后跳转到 GitHub 的授权页面的结果

图 4-5　授权后跳转到 127.0.0.1:7073 的结果

图 4-6　在浏览器中输入 127.0.0.1:7073/hi 后的结果

4.3　JWT 的简单应用

视频讲解

JWT(JSON Web Token)将用户信息加密到令牌(Token)里,服务器不保存任何用户信息。服务器使用自己保存的密钥来验证令牌(Token)的正确性,只要令牌(Token)正确就通过验证。

4.3.1　创建项目并添加依赖

用 IDEA 创建完项目 jwtexample 之后,确保在文件 pom.xml 的< dependencies >和</dependencies >之间添加了 Security、JWT、Web、Lang3、Guava 依赖,代码如例 4-12 所示。

【例 4-12】　添加 Security、JWT、Web、Lang3、Guava 依赖的代码示例。

```
< dependency >
    < groupId > org.springframework.cloud </groupId >
    < artifactId > spring - cloud - starter - security </artifactId >
</dependency >
< dependency >
    < groupId > io.jsonwebtoken </groupId >
    < artifactId > jjwt </artifactId >
    < version > 0.9.0 </version >
</dependency >
< dependency >
    < groupId > org.springframework.boot </groupId >
```

```xml
            <artifactId>spring-boot-starter-web</artifactId>
</dependency>
<dependency>
            <groupId>org.apache.commons</groupId>
            <artifactId>commons-lang3</artifactId>
            <version>3.8</version>
</dependency>
<dependency>
            <groupId>com.google.guava</groupId>
            <artifactId>guava</artifactId>
            <version>24.0-jre</version>
</dependency>
```

4.3.2　创建类 User

在包 com.bookcode 中创建 entity 子包，并在包 com.bookcode.entity 中创建类 User，代码如例 4-13 所示。

【例 4-13】　创建类 User 的代码示例。

```java
package com.bookcode.entity;
import java.util.List;
public class User {
    private Long id;
    private List<String> roles;
    private String username;
    public Long getId() {
        return id;
    }
    public void setId(Long id) {
        this.id = id;
    }
    public List<String> getRoles() {
        return roles;
    }
    public void setRoles(List<String> roles) {
        this.roles = roles;
    }
    public String getUsername() {
        return username;
    }
    public void setUsername(String username) {
        this.username = username;
    }
}
```

4.3.3 创建类 TokenUserAuthentication

在包 com.bookcode.entity 中创建类 TokenUserAuthentication,代码如例 4-14 所示。

【例 4-14】 创建类 TokenUserAuthentication 的代码示例。

```java
package com.bookcode.entity;
import org.springframework.security.core.Authentication;
import org.springframework.security.core.GrantedAuthority;
import org.springframework.security.core.authority.SimpleGrantedAuthority;
import java.util.Collection;
import java.util.stream.Collectors;
//Spring Security 中存放认证用户的信息
public class TokenUserAuthentication implements Authentication {
    private static final long serialVersionUID = 3730332217518791533L;
    private User user;
    private Boolean authentication = false;
    public TokenUserAuthentication(User user , Boolean authentication) {
        this.user = user ;
        this.authentication = authentication;
    }
    @Override
    public Collection<? extends GrantedAuthority> getAuthorities() {
        return user.getRoles().stream()
            .map(SimpleGrantedAuthority::new).collect(Collectors.toList());
    }
    @Override
    public Object getCredentials() {
        return "";
    }
    @Override
    public Object getDetails() {
        return user;
    }
    @Override
    public Object getPrincipal() {
        return user.getUsername();
    }
    @Override
    public boolean isAuthenticated() {
        return authentication;
    }
    @Override
    public void setAuthenticated(boolean isAuthenticated) throws IllegalArgumentException {
        this.authentication = isAuthenticated;
```

```
    }
    @Override
    public String getName() {
        return user.getUsername();
    }
}
```

4.3.4 创建类 JwtUtil

在包 com.bookcode 中创建 util 子包,并在包 com.bookcode.util 中创建类 JwtUtil,代码如例 4-15 所示。

【例 4-15】 创建类 JwtUtil 的代码示例。

```
package com.bookcode.util;
import com.bookcode.entity.TokenUserAuthentication;
import com.bookcode.entity.User;
import io.jsonwebtoken.Claims;
import io.jsonwebtoken.Jwts;
import io.jsonwebtoken.SignatureAlgorithm;
import org.apache.commons.lang3.math.NumberUtils;
import org.springframework.beans.factory.annotation.Value;
import org.springframework.security.core.Authentication;
import javax.servlet.http.HttpServletRequest;
import java.util.*;
public class JwtUtil {
    //Token 过期时间设置为 24 小时
    private static final long VALIDITY_TIME_MS = 24 * 60 * 60 * 1000;
    //header 中标识
    private static final String AUTH_HEADER_NAME = "x-authorization";
    static final String OtherSECRET = "ThisIsASecret";
    //签名密钥
    @Value("${jwt.token.secret}")
    private String secret;
    public static void validateToken(String token) {
        try {
            //解析 Token
            Map<String, Object> body = Jwts.parser()
                    .setSigningKey(OtherSECRET)
                    .parseClaimsJws(token.replace("Bearer ",""))
                    .getBody();
        }catch (Exception e){
            throw new IllegalStateException("Invalid Token. " + e.getMessage());
        }
    }
```

```java
//验签
public Optional<Authentication> verifyToken(HttpServletRequest request) {
    final String token = request.getHeader(AUTH_HEADER_NAME);
    if (token != null && !token.isEmpty()){
        final User user = parse(token.trim());
        if (user != null) {
            return Optional.of(new TokenUserAuthentication(user, true));
        }
    }
    return Optional.empty();
}
//为用户创建一个 JWT 的 Token
public String create(User user) {
    return Jwts.builder()
            .setExpiration(new Date(System.currentTimeMillis() + VALIDITY_TIME_MS))
            .setSubject(user.getUsername())
            .claim("id", user.getId())
            .claim("roles", user.getRoles())
            .signWith(SignatureAlgorithm.HS256, secret)
            .compact();
}
//从 Token 中取出用户信息
public User parse(String token) {
    Claims claims = Jwts.parser()
            .setSigningKey(secret)
            .parseClaimsJws(token)
            .getBody();
    User user = new User();
    user.setId( NumberUtils.toLong(claims.getId()));
    user.setUsername(claims.get("username",String.class));
    user.setRoles((List<String>) claims.get("roles"));
    return user;
}
}
```

4.3.5 创建类 HelloController

在包 com.bookcode 中创建 controller 子包，并在包 com.bookcode.controller 中创建类 HelloController，代码如例 4-16 所示。

【例 4-16】 创建类 HelloController 的代码示例。

```java
package com.bookcode.controller;
import com.bookcode.util.JwtUtil;
import com.bookcode.entity.User;
```

```java
import com.google.common.collect.Lists;
import io.jsonwebtoken.SignatureAlgorithm;
import org.apache.commons.lang3.StringUtils;
import org.springframework.web.bind.annotation.*;
import io.jsonwebtoken.Jwts;
import javax.annotation.Resource;
import java.util.Date;
import java.util.HashMap;
import java.util.Map;
@RestController
public class HelloController {
    @Resource
    private JwtUtil jwtTokenUtil;
    static final String OtherSECRET = "ThisIsASecret";
    //该方法获取 Token
    @GetMapping("/test")
    public Map login(String username, String password) {
        Map<String, Object> map = new HashMap<>();
        if (!StringUtils.equals( username, "demo") || !StringUtils.equals(password, "demo")) {
            map.put("status", 4);
            map.put("msg", "登录失败,用户名密码错误");
            return map;
        }
        User user = new User();
        user.setUsername(username);
        user.setRoles(Lists.newArrayList("ROLE_ADMIN"));
        user.setId(1L);
        map.put("状态", 1);
        map.put("信息", "登录成功");
        map.put("token", jwtTokenUtil.create(user));
        return map;
    }
    @GetMapping("/hello")
    public @ResponseBody
    Object hellWorld() {
        return "Hello World! This is a JWT example.";
    }
}
    @GetMapping("/returntoken")
    public static String returntoken (String username){
        HashMap<String, Object> map = new HashMap<>();
        map.put("username", username);
        String jwt = Jwts.builder()
                .setClaims(map)
                .setExpiration(new Date(System.currentTimeMillis() + 3600_000_000L))// 1000 hour
                .signWith(SignatureAlgorithm.HS512, OtherSECRET)
                .compact();
        return "JWT是: \n Bearer " + jwt; //jwt 加 Bearer
    }
}
```

4.3.6 创建文件 index.html

在目录 src/main/resources/static 下,创建文件 index.html,代码如例 4-17 所示。

【例 4-17】 创建文件 index.html 的代码示例。

```html
<!DOCTYPE html>
<html lang="en">
<head>
    <meta charset="UTF-8">
    <title>JWT Spring Security Demo</title>
    <link rel="stylesheet" href="https://maxcdn.bootstrapcdn.com/bootstrap/3.3.7/css/bootstrap.min.css">
    <link rel="stylesheet" href="https://maxcdn.bootstrapcdn.com/bootstrap/3.3.7/css/bootstrap-theme.min.css">
    <script src="https://code.jquery.com/jquery-2.2.2.js"></script>
    <script>
    $.ajaxSetup({
      contentType: "application/json; charset=utf-8"
    });
    var token = "";
    $(document).ready(function(){
        $("#exampleServiceBtn").click(function(){
            $.ajax({
                url: '/hello',
                headers: {'Authorization': token},
                method: 'GET'
            }).always(function(data, status, xhr) {
                if(data.responseJSON)
                    $("#response").text(JSON.stringify(data.responseJSON, null, 4));
                else
                    $("#response").text(JSON.stringify(data));
            });
        });
    });
    </script>
</head>
<body>
<div class="container">
    <h1>Spring Cloud JWT Demo</h1>
        <div class="col-md-6">
<button type="button" class="btn btn-default" id="exampleServiceBtn" style="margin-bottom: 16px;">
                call example service
            </button>
```

```html
                <div class = "panel panel-default">
                    <div class = "panel-heading">
                        <h3 class = "panel-title">Response:</h3>
                    </div>
                    <div class = "panel-body">
                        <pre id = "response"></pre>
                    </div>
                </div>
            </div>
            <div class = "col-md-6">
            </div>
            <div class = "col-md-6">
            <div id = "logout" style = "display:none">
                <div class = "panel panel-default">
                    <div class = "panel-heading">
                        <h3 class = "panel-title">Authenticated user</h3>
                        Spring Cloud Gateway</div>
                    <div class = "panel-body">
                        <div id = "userInfoBody"></div>
                        <button id = "btn-logout" type = "button" class = "btn btn-default">logout</button>
                    </div>
                </div>
            </div>
        </div>
    </div>
</body>
</hrml>
```

4.3.7 创建配置文件 application.yml

在目录 src/main/resources 下,创建配置文件 application.yml,代码如例 4-18 所示。

【例 4-18】 创建的配置文件 application.yml 代码示例。

```yaml
spring:
  security:
    basic:
      enabled: true          # 开启基于 HTTP basic 的认证
    user:
      name: zs               # 配置登录的账号
      password: zs           # 配置登录的密码
logging:
```

```
    level:
      root: info
      org :
        springframework:
          security: DEBUG
jwt:
  token:
    secret : ThisIsASecret
```

4.3.8 修改入口类

修改入口类,修改后的代码如例 4-19 所示。

【例 4-19】 修改后的入口类的代码示例。

```
package com.bookcode;
import com.bookcode.util.JwtUtil;
import org.springframework.boot.SpringApplication;
import org.springframework.boot.autoconfigure.SpringBootApplication;
import org.springframework.context.annotation.Bean;
@SpringBootApplication
public class DemoApplication {
    @Bean
    public JwtUtil configToken() {
      return new JwtUtil();
    }
    public static void main(String[] args) {
        SpringApplication.run(DemoApplication.class, args);
    }
}
```

4.3.9 运行程序

运行程序后,在浏览器中输入 localhost:8080/hello,页面跳转到 localhost:8080/login,结果如图 4-7 所示。输入正确的 User 和 Password 后,页面跳转到 localhost:8080/hello,结果如图 4-8 所示。在浏览器中输入 localhost:8080/test? username=demo&password=demo,结果如图 4-9 所示。在浏览器中输入 localhost:8080/returntoken,结果如图 4-10 所示。在浏览器中输入 localhost:8080/index.html,结果如图 4-11 所示。

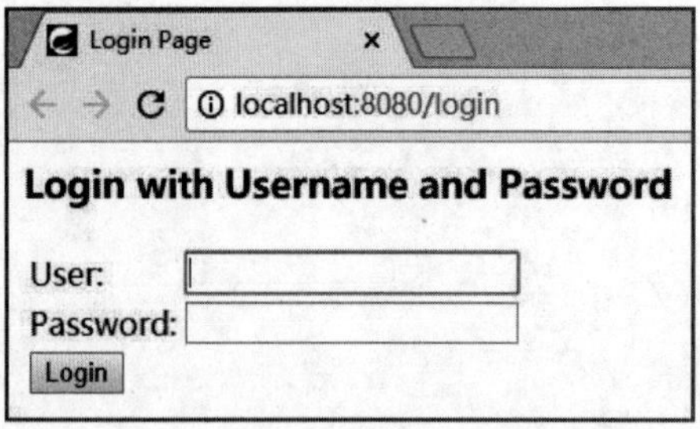

图 4-7 在浏览器中输入 localhost:8080/hello 后页面跳转到 localhost:8080/login 的结果

图 4-8 输入正确的 User 和 Password 后页面跳转到 localhost:8080/hello 的结果

图 4-9 在浏览器中输入 localhost:8080/test? username=demo&password=demo 后的结果

图 4-10 在浏览器中输入 localhost:8080/returntoken 的结果

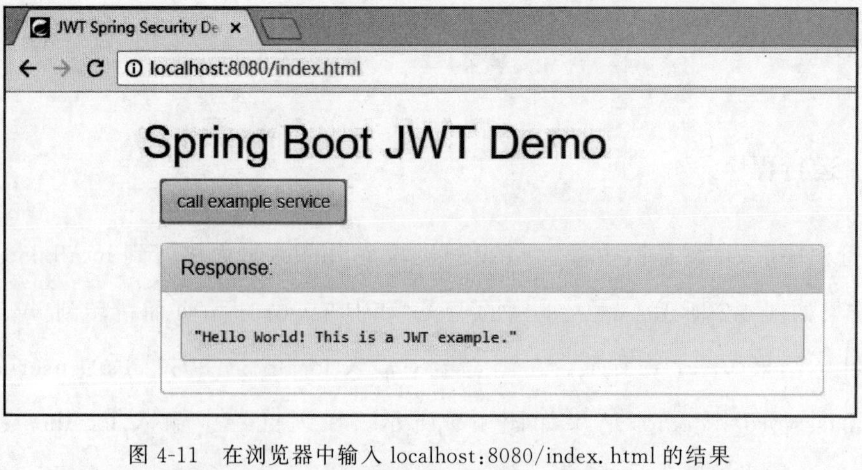

图 4-11 在浏览器中输入 localhost:8080/index.html 的结果

4.4 Gateway、JWT、Actuator 的综合应用

视频讲解

本节介绍 Spring Cloud Gateway、JWT 和 Actuator 的综合应用。Actuator 主要是完成微服务的监控和治理。查看微服务间的数据处理和调用,当它们之间出现了异常时,就可以快速地定位到出现问题的地方。

4.4.1 创建项目并添加依赖

用 IDEA 创建完项目 gateway-server 之后,确保在文件 pom.xml 的<dependencies>和</dependencies>之间添加了 JWT、Web、Gateway、Actuator 依赖,代码如例 4-20 所示。

【例 4-20】 添加 JWT、Web、Gateway、Actuator 依赖的代码示例。

```
<dependency>
        <groupId>io.jsonwebtoken</groupId>
        <artifactId>jjwt</artifactId>
        <version>0.9.0</version>
</dependency>
<dependency>
        <groupId>org.springframework.boot</groupId>
        <artifactId>spring-boot-starter-web</artifactId>
</dependency>
<dependency>
        <groupId>org.springframework.cloud</groupId>
        <artifactId>spring-cloud-starter-gateway</artifactId>
</dependency>
<dependency>
        <groupId>org.springframework.boot</groupId>
        <artifactId>spring-boot-starter-actuator</artifactId>
</dependency>
```

4.4.2 创建类 JwtUtil

在包 com.bookcode 中创建 util 子包,并在包 com.bookcode.util 中创建类 JwtUtil,代码如例 4-21 所示。

【例 4-21】 创建类 JwtUtil 的代码示例。

```
package com.bookcode.util;
import java.util.HashMap;
```

```java
import java.util.Random;
import io.jsonwebtoken.Jwts;
import io.jsonwebtoken.SignatureAlgorithm;
public class JwtUtil {
public static final String SECRET = "qazwsx123444 $ # % # ( ) * && asdaswwi1235 ?;!@ #
kmmmpom in * * * xx * * &";
    public static final String TOKEN_PREFIX = "Bearer";
    public static String generateToken(String user) {
        HashMap<String, Object> map = new HashMap<>();
        map.put("id", new Random().nextInt());
        map.put("user", user);
        String jwt = Jwts.builder()
                .setSubject("user info").setClaims(map)
                .signWith(SignatureAlgorithm.HS512, SECRET)
                .compact();
        String finalJwt = TOKEN_PREFIX + " " + jwt;
        return finalJwt;
    }
}
```

4.4.3 创建类 HelloController

在包 com.bookcode 中创建 controller 子包,并在包 com.bookcode.controller 中创建类 HelloController,代码如例 4-22 所示。

【例 4-22】 创建类 HelloController 的代码示例。

```java
package com.bookcode.controller;
import com.bookcode.util.JwtUtil;
import org.springframework.web.bind.annotation.GetMapping;
import org.springframework.web.bind.annotation.PathVariable;
import org.springframework.web.bind.annotation.RequestMapping;
import org.springframework.web.bind.annotation.RestController;
import javax.servlet.http.HttpServletRequest;
import java.util.Enumeration;
@RestController
public class HelloController {
    @GetMapping("/getToken/{name}")
    public String get(@PathVariable("name") String name) {
        String jwt = "不存在,请检查用户名.";
        if(name.equals("admin") || name.equals("spring") || name.equals("cloud")) {
            jwt = JwtUtil.generateToken(name);
            return "用户" + name +"的 Token 为: " + jwt;
        }
        else {
```

```
            return "用户" + name + jwt;
        }
    }
    @RequestMapping("test")
    public String clienttest(HttpServletRequest request) {
        System.out.println("------------ 成功访问 test 方法! ------------");
        Enumeration headerNames = request.getHeaderNames();
        while (headerNames.hasMoreElements()) {
            String key = (String) headerNames.nextElement();
            System.out.println(key + ":" + request.getHeader(key));
        }
        return "成功访问 test 方法!";
    }
}
```

4.4.4 创建配置文件 application.yml

在目录 src/main/resources 下,创建配置文件 application.yml,代码如例 4-23 所示。

【例 4-23】 创建的配置文件 application.yml 代码示例。

```
info:
  app:
    name: gateway-server
    encoding: utf-8
    java: 1.8
```

4.4.5 修改入口类

修改入口类,修改后的代码如例 4-24 所示。

【例 4-24】 修改后的入口类的代码示例。

```
package com.bookcode;
import org.springframework.boot.SpringApplication;
import org.springframework.boot.autoconfigure.SpringBootApplication;
import org.springframework.context.annotation.Bean;
import org.springframework.http.codec.ServerCodecConfigurer;
@SpringBootApplication
public class DemoApplication {
    @Bean
```

```
    public ServerCodecConfigurer serverCodecConfigurer() {
        return ServerCodecConfigurer.create();
    }
    public static void main(String[] args) {
        SpringApplication.run(DemoApplication.class, args);
    }
}
```

4.4.6 运行程序

运行程序后,在浏览器中输入 localhost:8080/actuator/health,结果如图 4-12 所示。在浏览器中输入 localhost:8080/actuator/info,结果如图 4-13 所示。在浏览器中输入 localhost:8080/test,在浏览器中的结果如图 4-14 所示,在控制台中的结果如图 4-15 所示。在浏览器中输入 localhost:8080/getToken/admin,输入正确的用户名 admin,得到 admin 的 Token,结果如图 4-16 所示。在浏览器中输入 localhost:8080/getToken/zs,输入的用户名不正确,结果如图 4-17 所示。

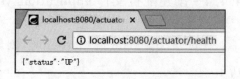

图 4-12　在浏览器中输入 localhost:8080/actuator/health 后的结果

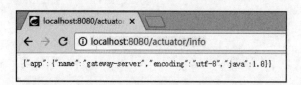

图 4-13　在浏览器中输入 localhost:8080/actuator/info 后的结果

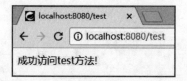

图 4-14　在浏览器中输入 localhost:8080/test 后浏览器中的结果

```
-----------------成功访问test方法!-----------------
host: localhost:8080
connection: keep-alive
upgrade-insecure-requests: 1
user-agent: Mozilla/5.0 (Windows NT 10.0; Win64; x64) AppleWebKit/537.36 (KHTML, like Gecko)
accept: text/html,application/xhtml+xml,application/xml;q=0.9,image/webp,image/apng,*/*;q=0.8
accept-encoding: gzip, deflate, br
accept-language: zh-CN,zh;q=0.9,en;q=0.8
```

图 4-15　在浏览器中输入 localhost:8080/test 后控制台中的结果

图 4-16 在浏览器中输入 localhost:8080/getToken/admin 的结果

图 4-17 在浏览器中输入 localhost:8080/getToken/zs 的结果

4.5 Eureka、Zuul、OAuth2 和 JWT 的综合应用

视频讲解

4.5.1 zuul-server 的实现

用 IDEA 创建完项目 zuul-server 之后,确保在文件 pom.xml 的< dependencies >和</dependencies >之间添加了 Zuul、Eureka Client、Security 和 OAuth2 依赖,代码如例 4-25 所示。

【例 4-25】 添加 Zuul、Eureka Client、Security 和 OAuth2 依赖的代码示例。

```
<dependency>
    <groupId>org.springframework.cloud</groupId>
    <artifactId>spring-cloud-starter-netflix-zuul</artifactId>
</dependency>
<dependency>
    <groupId>org.springframework.cloud</groupId>
    <artifactId>spring-cloud-starter-netflix-eureka-client</artifactId>
</dependency>
<dependency>
    <groupId>org.springframework.cloud</groupId>
    <artifactId>spring-cloud-starter-security</artifactId>
</dependency>
<dependency>
    <groupId>org.springframework.cloud</groupId>
    <artifactId>spring-cloud-starter-oauth2</artifactId>
</dependency>
```

在包 com.bookcode 中创建 controller 子包,并在包 com.bookcode.controller 中创建类 GreetController,代码如例 4-26 所示。

【例 4-26】 创建类 GreetController 的代码示例。

```
package com.bookcode.controller;
import org.springframework.web.bind.annotation.GetMapping;
import org.springframework.web.bind.annotation.RestController;
@RestController
public class GreetController {
    @GetMapping("/")
    public String greet() {
        return "Hi, I am Zuul Server at Port 5555.";
    }
}
```

在目录 src/main/resources 下,创建配置文件 bootstrap.yml,代码如例 4-27 所示。一般在创建配置文件 bootstrap.yml 以后,都需要修改代码,后面的章节将创建并修改代码的过程简称为创建配置文件 bootstrap.yml。bootstrap.yml 先于 application.yml 加载,用于程序引导时执行。

【例 4-27】 创建的配置文件 bootstrap.yml 代码示例。

```
spring:
  application:
    name: zuul-server
server:
  port: 5555
eureka:
  client:
    serviceUrl:
      defaultZone: http://${eureka.host:127.0.0.1}:${eureka.port:8761}/eureka/
  instance:
    prefer-ip-address: true
zuul:
  routes:
    client-a:
      path: /client/**
      serviceId: client-a
security:
  basic:
    enabled: false
  oauth2:
    client:
      access-token-uri: http://localhost:7777/uaa/oauth/token         #令牌端点
      user-authorization-uri: http://localhost:7777/uaa/oauth/authorize  #授权端点
      client-id: zuul_server            #OAuth2 客户端 ID
      client-secret: secret             #OAuth2 客户端密钥
    resource:
      jwt:
        key-value: springcloud123       #使用对称加密方式,默认算法为 HS256
```

修改入口类,修改后的代码如例 4-28 所示。

【例 4-28】 修改后的入口类的代码示例。

```
package com.bookcode;
import org.springframework.boot.SpringApplication;
import org.springframework.boot.autoconfigure.SpringBootApplication;
import org.springframework.cloud.client.discovery.EnableDiscoveryClient;
import org.springframework.cloud.netflix.zuul.EnableZuulProxy;
import org.springframework.boot.autoconfigure.security.oauth2.client.EnableOAuth2Sso;
import org.springframework.security.config.annotation.web.builders.HttpSecurity;
import org.springframework.security.config.annotation.web.configuration.
WebSecurityConfigurerAdapter;
@EnableDiscoveryClient
@EnableZuulProxy
@EnableOAuth2Sso
@SpringBootApplication
public class DemoApplication extends WebSecurityConfigurerAdapter {
    @Override
    protected void configure(HttpSecurity http) throws Exception {
        http
            .authorizeRequests()
            .antMatchers("/login", "/client/**")
            .permitAll()
            .anyRequest()
            .authenticated()
            .and()
            .csrf()
            .disable();
    }
    public static void main(String[] args) {
        SpringApplication.run(DemoApplication.class, args);
    }
}
```

4.5.2 auth-server 的实现

用 IDEA 创建完项目 auth-server 之后，确保在文件 pom.xml 的< dependencies >和</dependencies >之间添加了 Eureka Client 和 OAuth2 依赖，代码如例 4-29 所示。

【例 4-29】 添加 Eureka Client 和 OAuth2 依赖的代码示例。

```
<dependency>
        <groupId>org.springframework.cloud</groupId>
        <artifactId>spring-cloud-starter-netflix-eureka-client</artifactId>
</dependency>
<dependency>
        <groupId>org.springframework.cloud</groupId>
        <artifactId>spring-cloud-starter-oauth2</artifactId>
</dependency>
```

在包 com.bookcode 中创建 controller 子包,并在包 com.bookcode.controller 中创建类 HiController,代码如例 4-30 所示。

【例 4-30】 创建类 HiController 的代码示例。

```
package com.bookcode.controller;
import org.springframework.web.bind.annotation.GetMapping;
import org.springframework.web.bind.annotation.RestController;
@RestController
public class HiController {
    @GetMapping("/")
    public String welcome() {
        return "欢迎,我是 7777.";
    }
    @GetMapping("/hi")
    public String hi() {
        return "Hi, I am OAuth Server at Port 7777.";
    }
}
```

在包 com.bookcode 中创建类 OAuthConfiguration,代码如例 4-31 所示。

【例 4-31】 创建类 OAuthConfiguration 的代码示例。

```
package com.bookcode;
import org.springframework.beans.factory.annotation.Autowired;
import org.springframework.context.annotation.Bean;
import org.springframework.context.annotation.Configuration;
import org.springframework.security.authentication.AuthenticationManager;
import org.springframework.security.oauth2.config.annotation.configurers.ClientDetailsServiceConfigurer;
import org.springframework.security.oauth2.config.annotation.web.configuration.AuthorizationServerConfigurerAdapter;
import org.springframework.security.oauth2.config.annotation.web.configuration.EnableAuthorizationServer;
import org.springframework.security.oauth2.config.annotation.web.configurers.AuthorizationServerEndpointsConfigurer;
import org.springframework.security.oauth2.provider.token.TokenStore;
import org.springframework.security.oauth2.provider.token.store.JwtAccessTokenConverter;
import org.springframework.security.oauth2.provider.token.store.JwtTokenStore;
@Configuration
@EnableAuthorizationServer
public class OAuthConfiguration extends AuthorizationServerConfigurerAdapter {
    @Autowired
    private AuthenticationManager authenticationManager;
    @Override
```

```java
        public void configure(ClientDetailsServiceConfigurer clients) throws Exception {
            clients
            .inMemory()
            .withClient("zuul_server")
            .secret("secret")
            .scopes("WRIGTH", "read").autoApprove(true)
            .authorities("WRIGTH_READ", "WRIGTH_WRITE")
            .authorizedGrantTypes("implicit", "refresh_token", "password", "authorization_code");
        }
        @Override
        public void configure ( AuthorizationServerEndpointsConfigurer endpoints) throws Exception {
            endpoints
            .tokenStore(jwtTokenStore())
            .tokenEnhancer(jwtTokenConverter())
            .authenticationManager(authenticationManager);
        }
        @Bean
        public TokenStore jwtTokenStore() {
            return new JwtTokenStore(jwtTokenConverter());
        }
        @Bean
        protected JwtAccessTokenConverter jwtTokenConverter() {
            JwtAccessTokenConverter converter = new JwtAccessTokenConverter();
            converter.setSigningKey("springcloud123");
            return converter;
        }
}
```

在目录 src/main/resources 下,创建配置文件 bootstrap.yml,代码如例 4-32 所示。

【例 4-32】 创建的配置文件 bootstrap.yml 代码示例。

```yml
spring:
  application:
    name: auth-server
server:
  port: 7777
  servlet:
    contextPath: /uaa      #Web 基路径
eureka:
  client:
    serviceUrl:
      defaultZone: http://${eureka.host:127.0.0.1}:${eureka.port:8761}/eureka/
  instance:
    prefer-ip-address: true
```

修改入口类，修改后的代码如例 4-33 所示。

【例 4-33】 修改后的入口类的代码示例。

```java
package com.bookcode;
import org.springframework.boot.SpringApplication;
import org.springframework.boot.autoconfigure.SpringBootApplication;
import org.springframework.cloud.client.discovery.EnableDiscoveryClient;
import org.springframework.context.annotation.Bean;
import org.springframework.security.authentication.AuthenticationManager;
import org.springframework.security.config.BeanIds;
import org.springframework.security.config.annotation.authentication.builders.AuthenticationManagerBuilder;
import org.springframework.security.config.annotation.web.configuration.WebSecurityConfigurerAdapter;
import org.springframework.security.crypto.password.NoOpPasswordEncoder;
@SpringBootApplication
@EnableDiscoveryClient
public class DemoApplication extends WebSecurityConfigurerAdapter {
    public static void main(String[] args) {
        SpringApplication.run(DemoApplication.class, args);
    }
    @Bean(name = BeanIds.AUTHENTICATION_MANAGER)
    @Override
    public AuthenticationManager authenticationManagerBean() throws Exception {
        return super.authenticationManagerBean();
    }
    @Override
    protected void configure(AuthenticationManagerBuilder auth) throws Exception {
        auth
                .inMemoryAuthentication()
                .withUser("guest").password("guest").authorities("WRIGTH_READ")
                .and()
            .withUser("admin").password("admin").authorities("WRIGTH_READ", "WRIGTH_WRITE");
    }
    @Bean
    public static NoOpPasswordEncoder passwordEncoder() {
        return (NoOpPasswordEncoder) NoOpPasswordEncoder.getInstance();
    }
}
```

4.5.3 client-a 的实现

用 IDEA 创建完项目 client-a 之后，确保在文件 pom.xml 的< dependencies >和</dependencies >之间添加了 Eureka Client、Security 和 OAuth2 依赖，代码如例 4-34 所示。

【例 4-34】 添加 Eureka Client、Security 和 OAuth2 依赖的代码示例。

```xml
<dependency>
            <groupId>org.springframework.cloud</groupId>
            <artifactId>spring-cloud-starter-netflix-eureka-client</artifactId>
</dependency>
<dependency>
            <groupId>org.springframework.cloud</groupId>
            <artifactId>spring-cloud-starter-security</artifactId>
</dependency>
<dependency>
            <groupId>org.springframework.cloud</groupId>
            <artifactId>spring-cloud-starter-oauth2</artifactId>
</dependency>
```

在包 com.bookcode 中创建 controller 子包，并在包 com.bookcode.controller 中创建类 HelloController，代码如例 4-35 所示。

【例 4-35】 创建类 HelloController 的代码示例。

```java
package com.bookcode.controller;
import org.springframework.web.bind.annotation.GetMapping;
import org.springframework.web.bind.annotation.RequestMapping;
import org.springframework.web.bind.annotation.RestController;
import javax.servlet.http.HttpServletRequest;
import java.util.Enumeration;
@RestController
public class HelloController {
    @RequestMapping("/test")
    public String test(HttpServletRequest request) {
        System.out.println("---------------- header ----------------");
        Enumeration headerNames = request.getHeaderNames();
        while (headerNames.hasMoreElements()) {
            String key = (String) headerNames.nextElement();
            System.out.println(key + ": " + request.getHeader(key));
        }
        System.out.println("---------------- header ----------------");
        return "您好，您已经通过验证！";
    }
    @GetMapping("/hello")
    public String hello() {
        return "Hello, I am a client at Port 8788.";
    }
    @GetMapping("/add")
    public String add(Integer a, Integer b){
        System.out.println("进入 client-a!");
```

```
            Integer c = a + b;
            return a.toString() + "+" + b.toString() + "=" + c.toString();
        }
        @GetMapping("/sub")
        public String sub(Integer a, Integer b){
            System.out.println("进入client-a!");
            Integer c = a - b;
            return a.toString() + "-" + b.toString() + "=" + c.toString();
        }
        @GetMapping("/mul")
        public String mul(Integer a, Integer b){
            System.out.println("进入client-a!");
            Integer c = a * b;
            return a.toString() + "*" + b.toString() + "=" + c.toString();
        }
        @GetMapping("/div")
        public String div(Integer a, Integer b){
            System.out.println("进入client-a!");
            Integer c = a / b;
            return a.toString() + "/" + b.toString() + "=" + c.toString();
        }
    }
```

在目录 src/main/resources 下,创建配置文件 bootstrap.yml,代码如例 4-36 所示。

【例 4-36】 创建的配置文件 bootstrap.yml 代码示例。

```yaml
server:
  port: 8788
spring:
  application:
    name: client-a
eureka:
  client:
    serviceUrl:
      defaultZone: http://${eureka.host:127.0.0.1}:${eureka.port:8761}/eureka/
  instance:
    prefer-ip-address: true
```

修改入口类,修改后的代码如例 4-37 所示。

【例 4-37】 修改后的入口类的代码示例。

```
package com.bookcode;
import org.springframework.boot.SpringApplication;
```

```java
import org.springframework.boot.autoconfigure.SpringBootApplication;
import org.springframework.cloud.client.discovery.EnableDiscoveryClient;
import org.springframework.context.annotation.Bean;
import org.springframework.http.HttpMethod;
import org.springframework.security.config.annotation.web.builders.HttpSecurity;
import org.springframework.security.oauth2.config.annotation.web.configuration.
EnableResourceServer;
import org.springframework.security.oauth2.config.annotation.web.configuration.
ResourceServerConfigurerAdapter;
import org.springframework.security.oauth2.config.annotation.web.configurers.
ResourceServerSecurityConfigurer;
import org.springframework.security.oauth2.provider.token.TokenStore;
import org.springframework.security.oauth2.provider.token.store.JwtAccessTokenConverter;
import org.springframework.security.oauth2.provider.token.store.JwtTokenStore;
@SpringBootApplication
@EnableDiscoveryClient
@EnableResourceServer
public class DemoApplication extends ResourceServerConfigurerAdapter {
    public static void main(String[] args) {
        SpringApplication.run(DemoApplication.class, args);
    }
    @Override
    public void configure(HttpSecurity http) throws Exception {
        http
                .csrf().disable()
                .authorizeRequests()
                .antMatchers("/**").authenticated()
                .antMatchers(HttpMethod.GET, "/test")
                .hasAuthority("WRIGTH_READ");
    }
    @Override
    public void configure(ResourceServerSecurityConfigurer resources) throws Exception {
        resources
                .resourceId("WRIGTH")
                .tokenStore(jwtTokenStore());
    }
    @Bean
    protected JwtAccessTokenConverter jwtTokenConverter() {
        JwtAccessTokenConverter converter = new JwtAccessTokenConverter();
        converter.setSigningKey("springcloud123");
        return converter;
    }
    @Bean
    public TokenStore jwtTokenStore() {
        return new JwtTokenStore(jwtTokenConverter());
    }
}
```

4.5.4 运行程序

依次运行3.2节中eureka-server和本节的zuul-server、auth-server和client-a程序。运行程序后,首次在浏览器中输入 localhost:5555/client/test,会出现unauthorized(没有得到授权)的提示信息,结果如图4-18所示。在浏览器中输入localhost:5555,页面跳转到localhost:7777/uaa/login,结果如图4-19所示。输入正确的User和Password后(本例子中二者均为admin),页面跳转到localhost:5555,结果如图4-20所示。在浏览器中输入localhost:7777/uaa/,结果如图4-21所示。在浏览器中输入 localhost:7777/uaa/hi,结果如图 4-22 所示。再次在浏览器中输入localhost:5555/client/test,结果如图4-23所示。在浏览器中输入localhost:5555/client/hello,结果如图4-24所示。在浏览器中输入localhost:5555/client/add? a=3&b=2,浏览器中的结果如图4-25所示;控制台中的结果如图4-26所示,其中包括JWT等信息。

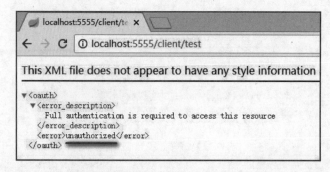

图4-18 首次在浏览器中输入localhost:5555/client/test的结果

图4-19 在浏览器中输入localhost:5555后页面　　　图4-20 输入正确User和Password后页面
　　　跳转到localhost:7777/uaa/login的结果　　　　　　　　跳转到localhost:5555的结果

localhost:7777/uaa/
localhost:7777/uaa/
欢迎，我是7777.

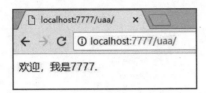

图 4-21 在浏览器中输入 localhost:7777/uaa/的结果

localhost:7777/uaa/hi
localhost:7777/uaa/hi
Hi, I am OAuth Server at Port 7777.

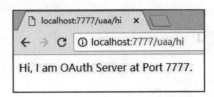

图 4-22 在浏览器中输入 localhost:7777/uaa/hi 的结果

localhost:5555/client/test
您好，您已经通过验证!

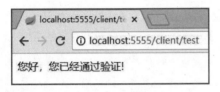

图 4-23 再次在浏览器中输入 localhost:5555/client/test 的结果

localhost:5555/client/hello
Hello, I am a client at Port 8788.

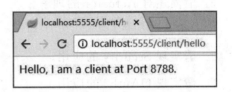

图 4-24 在浏览器中输入 localhost:5555/client/hello 的结果

localhost:5555/client/add?a=3&b=2
3+2=5

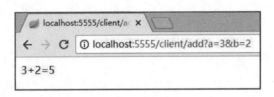

图 4-25 在浏览器中输入 localhost:5555/client/add?a=3&b=2 后浏览器中的结果

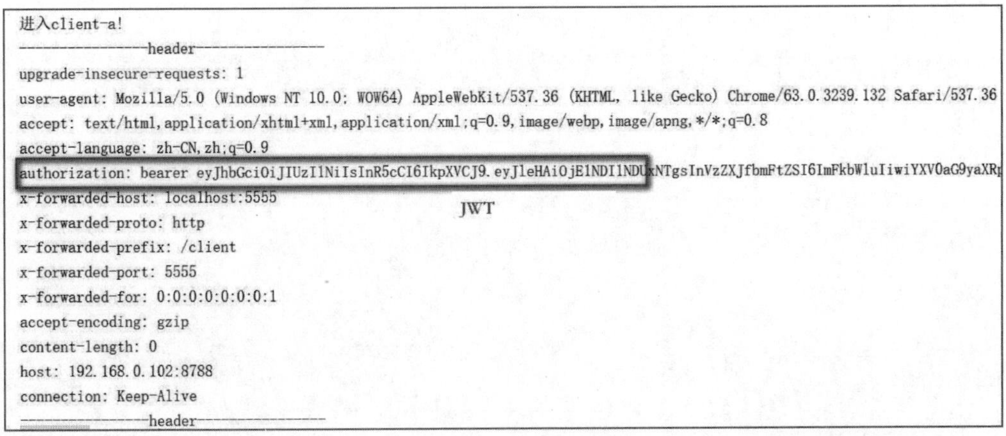

进入client-a!
---------------header---------------
upgrade-insecure-requests: 1
user-agent: Mozilla/5.0 (Windows NT 10.0; WOW64) AppleWebKit/537.36 (KHTML, like Gecko) Chrome/63.0.3239.132 Safari/537.36
accept: text/html,application/xhtml+xml,application/xml;q=0.9,image/webp,image/apng,*/*;q=0.8
accept-language: zh-CN,zh;q=0.9
authorization: bearer eyJhbGciOiJIUzI1NiIsInR5cCI6IkpXVCJ9.eyJleHAiOjE1NDI1NDUxNTgsInVzZXJfbmFtZSI6ImFkbWluIiwiYXV0aG9yaXR
JWT
x-forwarded-host: localhost:5555
x-forwarded-proto: http
x-forwarded-prefix: /client
x-forwarded-port: 5555
x-forwarded-for: 0:0:0:0:0:0:0:1
accept-encoding: gzip
content-length: 0
host: 192.168.0.102:8788
connection: Keep-Alive
---------------header---------------

图 4-26 在浏览器中输入 localhost:5555/client/add? a=3&b=2 后控制台中的结果

习题 4

一、问答题

1. 请简述对 OAuth 的理解。
2. 请简述对 JWT 的理解。
3. 请简述对 Actuator 的理解。

二、实验题

1. 请实现 Security 的应用。
2. 请实现 OAuth2 的应用。
3. 请实现 JWT 的应用。
4. 请实现 Security、OAuth2、JWT 等的综合应用。

第 5 章

Spring Cloud 断路器的应用

在微服务架构中，业务系统被拆分成细小的服务，服务与服务之间可以相互调用。由于网络原因或者自身的原因，服务并不能保证 100% 可用；如果单个服务出现问题，调用这个服务就会出现线程阻塞，此时若有大量的请求涌入，资源会被消耗完毕，导致服务瘫痪。由于服务之间的依赖性，故障会传播，从而会对整个系统造成灾难性的严重后果，这就是服务故障的"雪崩"效应。为了解决这个问题，业界提出了断路器模型。

本章介绍 Spring Cloud Hystrix 的应用、Spring Cloud Hystrix Dashboard 的应用和 Spring Cloud Turbine 的应用。

5.1 Spring Cloud Hystrix 的应用

视频讲解

互相依赖的服务之间不可避免地会发生调用失败，比如超时、异常等情况；如何保证在一个服务出现问题的情况下不会导致整个系统失败，这就是 Hystrix 需要做的事情。Hystrix 提供了熔断、隔离、回退、缓存、监控等功能，能够在一个或多个服务出现问题时保证系统依然可用。Netflix 开源了 Hystrix 组件，

Spring Cloud 对这一组件进行了整合，实现了断路器模式。

5.1.1 创建项目并添加依赖

用 IDEA 创建完项目 hystrixexample 之后，确保在文件 pom.xml 的<dependencies>和</dependencies>之间添加了 Hystrix、Eureka Client、Web、Ribbon、Openfeign 依赖，代码如例 5-1 所示。

【例 5-1】 添加 Hystrix、Eureka Client、Web、Ribbon、Openfeign 依赖的代码示例。

```xml
<dependency>
    <groupId>org.springframework.cloud</groupId>
    <artifactId>spring-cloud-starter-netflix-hystrix</artifactId>
</dependency>
<dependency>
    <groupId>org.springframework.cloud</groupId>
    <artifactId>spring-cloud-starter-netflix-eureka-client</artifactId>
</dependency>
<dependency>
    <groupId>org.springframework.boot</groupId>
    <artifactId>spring-boot-starter-web</artifactId>
</dependency>
<dependency>
    <groupId>org.springframework.cloud</groupId>
    <artifactId>spring-cloud-starter-netflix-ribbon</artifactId>
</dependency>
<dependency>
    <groupId>org.springframework.cloud</groupId>
    <artifactId>spring-cloud-starter-openfeign</artifactId>
</dependency>
```

5.1.2 创建接口 HiService

在包 com.bookcode 中创建 service 子包，并在包 com.bookcode.service 中创建接口 HiService，代码如例 5-2 所示。

【例 5-2】 创建接口 HiService 的代码示例。

```java
package com.bookcode.service;
import com.bookcode.controller.HiController;
import org.springframework.cloud.openfeign.FeignClient;
import org.springframework.web.bind.annotation.RequestMapping;
```

```
import org.springframework.web.bind.annotation.RequestMethod;
import org.springframework.web.bind.annotation.RequestParam;
@FeignClient(value = "service-hi",fallback = HiController.class)
public interface HiService {
    @RequestMapping(value = "/hi",method = RequestMethod.GET)
    String sayHiFromClientOne(@RequestParam(value = "name") String name);
}
```

5.1.3　创建类 HiController

在包 com.bookcode 中创建 controller 子包,并在包 com.bookcode.controller 中创建类 HiController,代码如例 5-3 所示。

【例 5-3】　创建类 HiController 的代码示例。

```
package com.bookcode.controller;
import com.bookcode.service.HiService;
import org.springframework.web.bind.annotation.RestController;
@RestController
public class HiController implements HiService {
    @Override
    public String sayHiFromClientOne(String name) {
        return name + ",您好!" + "hi 方法出错了!";
    }
}
```

5.1.4　创建类 HelloController

在包 com.bookcode.controller 中创建类 HelloController,代码如例 5-4 所示。

【例 5-4】　创建类 HelloController 的代码示例。

```
package com.bookcode.controller;
import com.netflix.hystrix.contrib.javanica.annotation.HystrixCommand;
import org.springframework.beans.factory.annotation.Autowired;
import org.springframework.web.bind.annotation.GetMapping;
import org.springframework.web.bind.annotation.RestController;
import org.springframework.web.client.RestTemplate;
@RestController
public class HelloController {
    @Autowired
    RestTemplate restTemplate;
//@HystrixCommand 注解方法创建了熔断器的功能,并指定了 fallbackMethod 熔断方法
```

```
        @HystrixCommand(fallbackMethod = "helloError")
        @GetMapping (value = "/hello")
        public String hello(String name) {
                return restTemplate.getForObject("http://localhost:8080/hi?name=" + name, String.class);
        }
        @HystrixCommand(fallbackMethod = "greetError")
        @GetMapping (value = "/greet")
        public String greet(String name) {
            return name + ", 欢迎您!";
        }
        public String helloError(String name) {
            return name + ",您好!" + "hello 方法出错了!";
        }
        public String greetError(String name) {
            return name + ",您好!" + "greet 方法出错了!";
        }
}
```

5.1.5 修改配置文件 application.properties

修改配置文件 application.properties，修改后的代码如例 5-5 所示。

【例 5-5】 修改后的配置文件 application.properties 的代码示例。

```
feign.hystrix.enabled = true
spring.application.name = hystrixexample
eureka.client.serviceUrl.defaultZone = http://localhost:8761/eureka/
info.app.name = hystrixexample
info.app.encoding = utf-8
info.app.java = 1.8
```

5.1.6 修改入口类

修改入口类，修改后的代码如例 5-6 所示。

【例 5-6】 修改后的入口类的代码示例。

```
package com.bookcode;
import org.springframework.boot.SpringApplication;
import org.springframework.boot.autoconfigure.SpringBootApplication;
import org.springframework.cloud.client.discovery.EnableDiscoveryClient;
import org.springframework.cloud.client.loadbalancer.LoadBalanced;
```

```
import org.springframework.cloud.netflix.hystrix.EnableHystrix;
import org.springframework.context.annotation.Bean;
import org.springframework.web.client.RestTemplate;
@EnableDiscoveryClient
@EnableHystrix
@SpringBootApplication
public class DemoApplication {
    @Bean
    @LoadBalanced
    RestTemplate restTemplate() {
        return new RestTemplate();
    }
    public static void main(String[] args) {
        SpringApplication.run(DemoApplication.class, args);
    }
}
```

5.1.7 运行程序

依次运行 3.2 节中 eureka-server 和本节的 hystrixexample 程序。运行程序后，在浏览器中输入 localhost:8080/actuator/health，结果如图 5-1 所示。在浏览器中输入 localhost:8080/actuator/info，结果如图 5-2 所示。在浏览器中输入 localhost:8080/hello?name=zs，结果如图 5-3 所示。在浏览器中输入 localhost:8080/hi?name=zs，结果如图 5-4 所示。在浏览器中输入 localhost:8080/greet?name=zs，结果如图 5-5 所示。

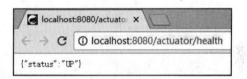

图 5-1 在浏览器中输入 localhost:8080/actuator/health 后的结果

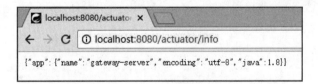

图 5-2 在浏览器中输入 localhost:8080/actuator/info 后的结果

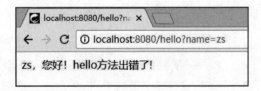

图 5-3 在浏览器中输入 localhost:8080/hello?name=zs 的结果

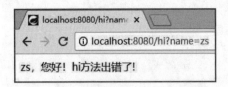

图 5-4　在浏览器中输入 localhost:8080/hi?name=zs 的结果

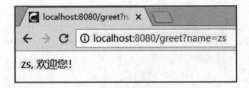

图 5-5　在浏览器中输入 localhost:8080/greet?name=zs 的结果

5.2　Spring Cloud Hystrix Dashboard 的应用

视频讲解

Spring Cloud Hystrix Dashboard 是可视化的监控工具。本节介绍 Spring Cloud Hystrix Dashboard 的应用。

5.2.1　添加依赖

在项目 hystrixexample 的基础，对其进行扩展。在文件 pom.xml 的<dependencies>和</dependencies>之间添加了 Spring Cloud Hystrix Dashboard 依赖，代码如例 5-7 所示。

【例 5-7】　添加 Spring Cloud Hystrix Dashboard 依赖的代码示例。

```xml
<!-->新添加的依赖<-->
<dependency>
        <groupId>org.springframework.cloud</groupId>
        <artifactId>spring-cloud-starter-netflix-hystrix-dashboard</artifactId>
</dependency>
```

5.2.2　修改入口类

修改入口类，修改后的代码如例 5-8 所示。

【例 5-8】　修改后的入口类的代码示例。

```java
package com.bookcode;
import com.netflix.hystrix.contrib.metrics.eventstream.HystrixMetricsStreamServlet;
import org.springframework.boot.SpringApplication;
import org.springframework.boot.autoconfigure.SpringBootApplication;
```

```java
import org.springframework.boot.web.servlet.ServletRegistrationBean;
import org.springframework.cloud.client.circuitbreaker.EnableCircuitBreaker;
import org.springframework.cloud.client.discovery.EnableDiscoveryClient;
import org.springframework.cloud.client.loadbalancer.LoadBalanced;
import org.springframework.cloud.netflix.hystrix.EnableHystrix;
import org.springframework.cloud.netflix.hystrix.dashboard.EnableHystrixDashboard;
import org.springframework.context.annotation.Bean;
import org.springframework.web.client.RestTemplate;
@EnableHystrixDashboard          //新增
@EnableCircuitBreaker            //新增
@EnableHystrix
@SpringBootApplication
public class DemoApplication {
    //下面新增两个 Bean 的代码
    @Bean
    public ServletRegistrationBean getServlet() {
        HystrixMetricsStreamServlet streamServlet = new HystrixMetricsStreamServlet();
        ServletRegistrationBean registrationBean = new ServletRegistrationBean(streamServlet);
        registrationBean.setLoadOnStartup(1);
        registrationBean.addUrlMappings("/hystrix.stream");
        registrationBean.setName("HystrixMetricsStreamServlet");
        return registrationBean;
    }
    @Bean
    @LoadBalanced
    RestTemplate restTemplate() {
        return new RestTemplate();
    }
    public static void main(String[] args) {
        SpringApplication.run(DemoApplication.class, args);
    }
}
```

5.2.3 运行程序

运行程序，在浏览器中输入 localhost:8080/greet?name＝zs 后，在浏览器中输入 localhost:8080/hystrix.stream，结果如图 5-6 所示。图 5-6 中是以文字形式显示运行状态。为了图形化和可视化，在浏览器中输入 localhost:8080/hystrix 后，在 URL 文本框中输入 localhost:8080/hystrix.stream，再为其命名（本例中的 afterGreetHystrixTest），如图 5-7 所示；单击 Monitor Stream 按钮，结果如图 5-8 所示。

图 5-6 在浏览器中输入 localhost:8080/hystrix.stream 的结果

图 5-7 在浏览器中输入 localhost:8080/hystrix 后输入 URL 并为其命名的结果

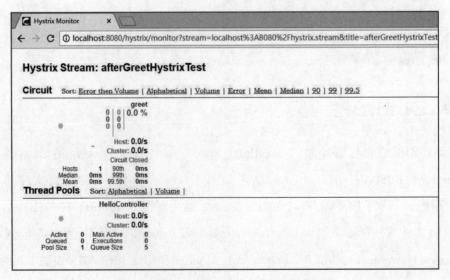

图 5-8 单击 Monitor Stream 按钮后的结果

5.3 Spring Cloud Turbine 的应用

视频讲解

5.3.1 创建项目并添加依赖

用 IDEA 创建完项目 turbineexample 之后,确保在文件 pom.xml 的<dependencies>和</dependencies>之间添加了 Hystrix、Turbine、Alctuator 依赖,代码如例 5-9 所示。

【例 5-9】 添加 Hystrix、Turbine、Alctuator 依赖的代码示例。

```
<dependency>
        <groupId>org.springframework.cloud</groupId>
        <artifactId>spring-cloud-starter-netflix-hystrix</artifactId>
</dependency>
<dependency>
        <groupId>org.springframework.cloud</groupId>
        <artifactId>spring-cloud-starter-turbine</artifactId>
</dependency>
<dependency>
        <groupId>org.springframework.cloud</groupId>
        <artifactId>spring-cloud-starter-netflix-turbine</artifactId>
</dependency>
<dependency>
        <groupId>org.springframework.boot</groupId>
        <artifactId>spring-boot-starter-actuator</artifactId>
</dependency>
```

5.3.2 修改配置文件 application.properties

修改配置文件 application.properties,修改后的代码如例 5-10 所示。

【例 5-10】 修改后的配置文件 application.properties 代码示例。

```
spring.application.name=service-turbine
server.port=8766
security.basic.enabled=false
eureka.client.serviceUrl.defaultZone=http://localhost:8761/eureka
turbine.aggregator.cluster-config = default
turbine.app-config = hystrixexample
turbine.cluster-name-expression = "default"
turbine.combine-host-port = true
turbine.instanceUrlSuffix = /hystrix.stream
```

5.3.3 修改入口类

修改入口类,修改后的代码如例 5-11 所示。

【例 5-11】 修改后的入口类的代码示例。

```
package com.bookcode;
import org.springframework.boot.SpringApplication;
import org.springframework.boot.autoconfigure.SpringBootApplication;
import org.springframework.cloud.netflix.turbine.EnableTurbine;
@EnableTurbine
@SpringBootApplication
public class DemoApplication {
    public static void main(String[] args) {
        SpringApplication.run(DemoApplication.class, args);
    }
}
```

5.3.4 运行程序

依次运行 3.2 节 eureka-server 程序和 5.1 节 hystrixexample 程序。

运行本节程序 turbineexample 后,在浏览器中输入 http://localhost:8766/turbine.stream,结果如图 5-8 所示。图 5-8 中是以文字形式显示运行状态。在浏览

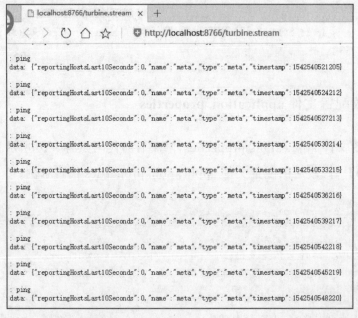

图 5-8 在浏览器中输入 http://localhost:8766/turbine.stream 的结果

器中输入 localhost:8080/greet? name=zs,结果如图 5-5 所示。为了图形化和可视化,在浏览器中输入 localhost:8080/hystrix,在图中 URL 文本框中输入 localhost:8766/turbine.stream,再为其命名(本例中的 turbine)并单击 Monitor Stream 按钮,结果如图 5-9 所示。

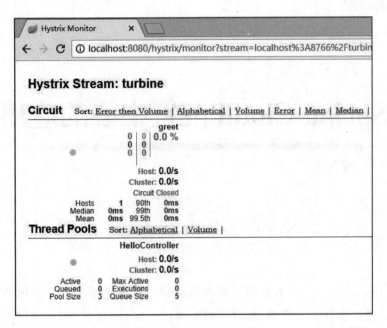

图 5-9 输入 URL 文本框后为其命名并单击 Monitor Stream 按钮结果

习题 5

一、问答题

请简述对断路器的理解。

二、实验题

1. 请实现 Spring Cloud Hystrix Dashboard 的应用。
2. 请实现 Spring Cloud Turbine 的应用。

第 6 章

Spring Cloud 配置中心的应用

在分布式系统中，一次请求的完成可能会调用很多个服务。由于服务数量巨大，为了方便对配置信息统一管理、部署、维护、实时更新，所以需要配置中心组件。

本章介绍如何实现 Spring Cloud Config Server、Spring Cloud Config Client、Spring Cloud Consul Configuration 和 Spring Cloud Zookeeper Configuration 等配置中心的应用。

6.1 Spring Cloud Config Server 的应用

视频讲解

在 Spring Cloud 中有分布式配置中心组件 Spring Cloud Config，它支持将服务配置放在本地，也支持放在远程配置仓库中（如 GitHub 仓库）。Spring Cloud Config 组件可以分为 Server 和 Client 两个角色；可以将配置文件集中放置在一个 GitHub 仓库里，再重新创建一个 Config Server 用于管理所有的配置文件；当需要更改配置时，只需要在本地更改后推送到远程仓库，所有的服务实例都可以通过 Config Server 获取配置文件，这时每个服务实例就相当于 Config Server 的 Config Client。为了保证系统的稳定，也可以对 Config Server 进行集群部署。

6.1.1 创建项目并添加依赖

用 IDEA 创建完项目 configserver 之后,确保在文件 pom.xml 的< dependencies >和</dependencies >之间添加了 Config Server 依赖,代码如例 6-1 所示。

【例 6-1】 添加 Config Server 依赖的代码示例。

```
< dependency >
        < groupId > org.springframework.cloud </groupId >
        < artifactId > spring - cloud - config - server </artifactId >
</dependency >
```

6.1.2 修改配置文件 application.properties

修改配置文件 application.properties,修改后的代码如例 6-2 所示。

【例 6-2】 修改后的配置文件 application.properties 代码示例。

```
spring.application.name = configserver
server.port = 8769
spring.cloud.config.server.git.uri = https://github.com/woodsheng/MicroserviceWithSpringCloud.git
spring.cloud.config.server.git.username =
spring.cloud.config.server.git.password =
```

6.1.3 修改入口类

修改入口类,修改后的代码如例 6-3 所示。

【例 6-3】 修改后的入口类的代码示例。

```
package com.bookcode;
import org.springframework.boot.SpringApplication;
import org.springframework.boot.autoconfigure.SpringBootApplication;
import org.springframework.cloud.config.server.EnableConfigServer;
@EnableConfigServer
@SpringBootApplication
public class DemoApplication {
    public static void main(String[] args) {
        SpringApplication.run(DemoApplication.class, args);
    }
}
```

6.1.4 运行程序

运行程序后,在浏览器中输入 localhost:8769/microservice-foo/dev,结果如图 6-1 所示。图 6-1 中的配置信息与图 6-2 中的在远程 GitHub 中的配置信息完全一致。

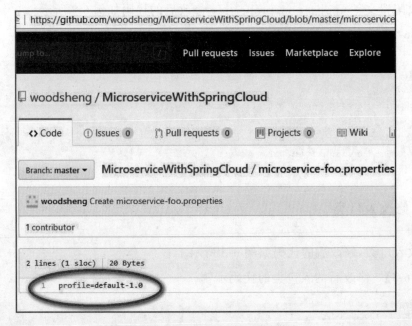

图 6-1 在浏览器中输入 localhost:8769/microservice-foo/dev 的结果

图 6-2 在远程 GitHub 中的配置信息

6.2 Spring Cloud Config Client 的应用

视频讲解

Spring Cloud Config Client 应用只需要在配置文件中增加要使用的 Config Server 的配置信息即可。本节介绍 Spring Cloud Config Client 的应用。

6.2.1 创建项目并添加依赖

用 IDEA 创建完项目 configclient 之后,确保在文件 pom.xml 的<dependencies>和</dependencies>之间添加了 Config Client、Web 依赖,代码如例 6-4 所示。

【例 6-4】 添加 Config Client、Web 依赖的代码示例。

```xml
<dependency>
        <groupId>org.springframework.cloud</groupId>
        <artifactId>spring-cloud-starter-config</artifactId>
</dependency>
<dependency>
        <groupId>org.springframework.boot</groupId>
        <artifactId>spring-boot-starter-web</artifactId>
</dependency>
```

6.2.2 创建类 HelloController

在包 com.bookcode 中创建 controller 子包,并在包 com.bookcode.controller 中创建类 HelloController,代码如例 6-5 所示。

【例 6-5】 创建类 HelloController 的代码示例。

```java
package com.bookcode.controller;
import org.springframework.beans.factory.annotation.Value;
import org.springframework.web.bind.annotation.RequestMapping;
import org.springframework.web.bind.annotation.RestController;
@RestController
public class HelloController {
    @Value("${profiles}")
    private String profiles;
    @RequestMapping(value = "/getProfiles")
    public String profiles() {
        return profiles;
    }
    @Value("${userName}")
    private String userName;
    @RequestMapping(value = "/getUserName")
    public String getUserName() {
        return "这是来自配置中心的信息,用户名为" + userName + ".";
    }
}
```

6.2.3 修改配置文件 application.properties

修改配置文件 application.properties，修改后的代码如例 6-6 所示。

【例 6-6】 修改后的配置文件 application.properties 代码示例。

```
spring.application.name = configclient
spring.cloud.config.uri = http://localhost:8769/ #指出 Config Server
server.port = 8768
spring.messages.encoding = UTF-8
profiles = 这是来自本地的配置信息。
```

6.2.4 运行程序

在运行 6.1 节 configserver 程序后，运行本节 configclient 程序。在浏览器中输入 localhost:8768/getUserName，结果如图 6-3 所示。在浏览器中输入 localhost:8768/getProfiles，结果如图 6-4 所示。

图 6-3　在浏览器中输入 localhost:8768/getUserName 的结果

图 6-4　在浏览器中输入 localhost:8768/getProfiles 的结果

6.3 Spring Cloud Consul 的应用

视频讲解

Spring Cloud Consul 提供了用于存储配置和其他元数据（键/值）对信息的功能。Spring Cloud Consul 是 Spring Cloud Config 的替代方案。当配置被加载到 Spring 环境中，默认情况下配置存储在文件夹 config 中。本节介绍 Spring Cloud Consul Configuration 的应用。

6.3.1 创建项目并添加依赖

用 IDEA 创建完项目 consulconfigexample 之后，确保在文件 pom.xml 的

<dependencies>和</dependencies>之间添加了 Consul、Web 依赖,代码如例 6-7 所示。

【例 6-7】 添加 Consul、Web 依赖的代码示例。

```xml
<dependency>
        <groupId>org.springframework.cloud</groupId>
        <artifactId>spring-cloud-starter-consul-all</artifactId>
        <!-- 包含了config,discover,bus依赖 -->
</dependency>
<dependency>
        <groupId>org.springframework.boot</groupId>
        <artifactId>spring-boot-starter-web</artifactId>
</dependency>
```

6.3.2 创建配置文件 application.yml

在目录 src/main/resources 下,创建配置文件 application.yml,代码如例 6-8 所示。

【例 6-8】 创建配置文件 application.yml 的代码示例。

```yaml
spring:
  cloud:
    consul:
      host: localhost
      port: 8500
      discovery:
        tags: foo=bar, baz
        healthCheckPath: /health
        healthCheckInterval: 15s
  application:
    name: consulconfigexample
server:
  port: 8880
```

6.3.3 创建配置文件 bootstrap.yml

在目录 src/main/resources 下,创建配置文件 bootstrap.yml,代码如例 6-9 所示。

【例 6-9】 创建配置文件 bootstrap.yml 的代码示例。

```yaml
spring:
  profiles:
    active: dev
```

```yaml
cloud:
  consul:
    config:
      #启用 consul config
      enabled: true
      #配置文件的前缀文件夹名
      prefix: config
      #共享配置的文件夹名
      defaultContext: share
      #profile 配置的文件夹名分隔符
      profileSeparator: '.'
```

6.3.4 修改入口类

修改入口类,修改后的代码如例 6-10 所示。

【例 6-10】 修改后的入口类的代码示例。

```java
package com.bookcode;
import org.springframework.boot.SpringApplication;
import org.springframework.boot.autoconfigure.SpringBootApplication;
import org.springframework.cloud.client.discovery.DiscoveryClient;
import org.springframework.cloud.client.discovery.EnableDiscoveryClient;
import org.springframework.web.bind.annotation.RequestMapping;
import org.springframework.web.bind.annotation.RestController;
import javax.annotation.Resource;
import org.springframework.beans.factory.annotation.Value;
import org.springframework.cloud.client.ServiceInstance;
import org.springframework.cloud.context.config.annotation.RefreshScope;
import java.util.List;
//consul 组件包含 eureka、config、bus 等 3 个组件的功能
@SpringBootApplication
@EnableDiscoveryClient        //注册服务和发现服务
@RestController
@RefreshScope                 //刷新配置项
public class DemoApplication {
    @Resource
    private DiscoveryClient discoveryClient;
    //注意,以下配置项需要在 consul 服务器上配置
    @Value("${firstname}")
    private String firstname;
    @Value("${year}")
    private Integer year;
    //从 consul 服务器获取服务实例信息的 JSON 格式数据
    @RequestMapping("/getInstances")
```

```
    public List<ServiceInstance> getInstances() {
        return discoveryClient.getInstances("consulconfigexample");
    }
//从 consul 服务器获取 Key/Value 配置
@RequestMapping("/keyValuePair")
public String keyValuePair() {
    return firstname + ":" + year;
}
```

6.3.5 运行程序

执行启动 Consul 服务器命令,如例 6-11 所示,启动 Consul 服务。

【例 6-11】 启动 Consul 服务的命令示例。

```
consul agent -dev
```

在浏览器中输入 localhost:8500;单击页面上方的超链接 Key/Value,单击 Create 按钮;输入键值对(config/consulconfigexample/firstname:张),如图 6-5 所示,单击 Save 按钮;输入键值对(config/consulconfigexample/year:21),如图 6-6 所示,单击 Save 按钮。

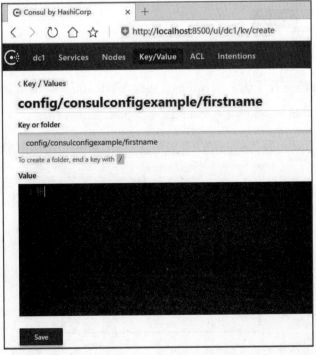

图 6-5　输入键值对(config/consulconfigexample/firstname:张)的结果

图 6-6 输入键值对(config/consulconfigexample/year:21)的结果

运行程序,在浏览器中输入 localhost:8880/getInstances,结果如图 6-7 所示。在浏览器中输入 localhost:8880/keyValuePair,结果如图 6-8 所示。

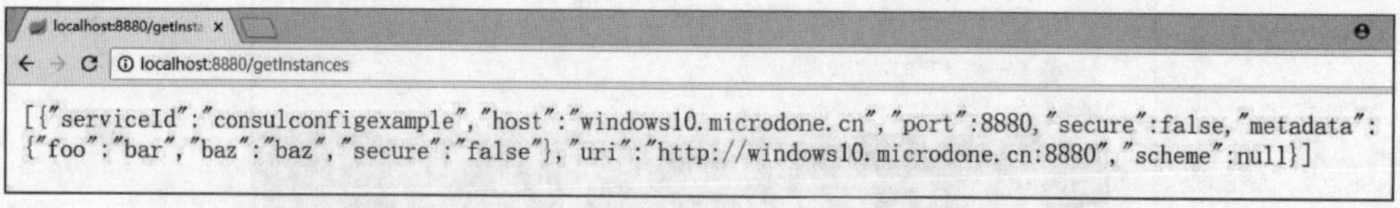

图 6-7 在浏览器中输入 localhost:8880/getInstances 的结果

图 6-8 在浏览器中输入 localhost:8880/keyValuePair 的结果

6.4　Spring Cloud Zookeeper 的应用

视频讲解

ZooKeeper 主要用于解决分布式集群中应用系统的一致性问题，它能提供基于类似于文件系统的目录节点树方式的数据存储。ZooKeeper 作用主要是用于维护和监控存储的数据的状态变化，通过监控这些数据状态的变化从而达到基于数据的集群管理。Spring Cloud Zookeeper 是 Spring Cloud Config 的替代方案。它通过 ZooKeeper 分级命名空间来储存配置项数据，还可以实时监听节点变化和通知机制。

6.4.1　创建项目并添加依赖

用 IDEA 创建完项目 zkconfigexample 之后，确保在文件 pom.xml 的<dependencies>和</dependencies>之间添加了 Zookeeper Configuration、Web 依赖，代码如例 6-12 所示。

【例 6-12】　添加 Zookeeper Configuration、Web 依赖的代码示例。

```xml
<dependency>
        <groupId>org.springframework.cloud</groupId>
        <artifactId>spring-cloud-starter-zookeeper-config</artifactId>
</dependency>
<dependency>
        <groupId>org.springframework.boot</groupId>
        <artifactId>spring-boot-starter-web</artifactId>
</dependency>
```

6.4.2　创建类 HelloController

在包 com.bookcode 中创建 controller 子包，并在包 com.bookcode.controller 中创建类 HelloController，代码如例 6-13 所示。

【例 6-13】　创建类 HelloController 的代码示例。

```java
package com.bookcode.controller;
import org.springframework.beans.factory.annotation.Value;
import org.springframework.web.bind.annotation.RequestMapping;
import org.springframework.web.bind.annotation.RestController;
@RestController
```

```
public class HelloController {
    @Value("${msg:defaultMsg}")
    private String msg;
    @RequestMapping(value = "/getMsg")
    public String getMsg() {
        return "这是来自配置中心的信息,用户名为" + msg + ".";
    }
}
```

6.4.3 创建配置文件 bootstrap.yml

在目录 src/main/resources 下,创建配置文件 bootstrap.yml,代码如例 6-14 所示。

【例 6-14】 创建的配置文件 bootstrap.yml 代码示例。

```
spring:
  application:
    name: zkconfigexample
  cloud:
    zookeeper:
      connect-string: localhost:2181
      config:
        enabled: false
server:
  port: 9106
```

6.4.4 运行程序

启动 ZooKeeper 服务器,命令如例 6-15 所示。

【例 6-15】 启动 ZooKeeper 服务器的命令示例。

```
zkServer
```

执行例 6-16 所示的命令,启动 ZooKeeper 客户端。

【例 6-16】 启动 ZooKeeper 客户端的命令示例。

```
zkCli
create /config ""
create /config/zkconfigexample ""
create /config/zkconfigexample/msg zhangsan
```

运行程序,在浏览器中输入 localhost:9106/getMsg,结果如图 6-9 所示。

图 6-9　在浏览器中输入 localhost:9106/getMsg 的结果

习题 6

实验题

1. 请实现基于 Spring Cloud Config 的配置应用开发。
2. 请实现基于 Spring Cloud Consul 的配置应用开发。
3. 请实现基于 Spring Cloud Zookeeper 的配置应用开发。

第7章

Spring Cloud 服务跟踪的应用

本章介绍如何实现 Spring Cloud Sleuth、Spring Cloud Zipkin 的应用开发。

7.1 Spring Cloud Sleuth 的应用

视频讲解

Spring Cloud Sleuth 是一个用于实现日志跟踪的强有力工具。Sleuth 可以应用于多线程服务或复杂的 Web 请求,尤其是在由多个服务组成的系统中。Sleuth 可以与日志框架 Logback、SLF4J 轻松地集成来跟踪和诊断问题。本节介绍 Spring Cloud Sleuth 的应用。

7.1.1 创建项目并添加依赖

用 IDEA 创建完项目 sleuthexample 之后,确保在文件 pom.xml 的< dependencies >和</ dependencies >之间添加了 Sleuth、Web 依赖,代码如例 7-1 所示。

【例 7-1】 添加 Sleuth、Web 依赖的代码示例。

```
< dependency >
    < groupId > org.springframework.cloud </ groupId >
```

```
            <artifactId>spring-cloud-starter-sleuth</artifactId>
</dependency>
<dependency>
            <groupId>org.springframework.boot</groupId>
            <artifactId>spring-boot-starter-web</artifactId>
</dependency>
```

7.1.2 创建类 SleuthService

在包 com.bookcode 中创建 service 子包,并在包 com.bookcode.service 中创建类 SleuthService,代码如例 7-2 所示。

【例 7-2】 创建类 SleuthService 的代码示例。

```
package com.bookcode.service;
import org.slf4j.Logger;
import org.slf4j.LoggerFactory;
import org.springframework.scheduling.annotation.Async;
import org.springframework.stereotype.Service;
@Service
public class SleuthService {
    private Logger logger = LoggerFactory.getLogger(this.getClass());
    public void doSomeWorkSameSpan() throws InterruptedException {
        Thread.sleep(1000L);
        logger.info("模拟在做一些事情.");
    }
    @Async
    public void asyncMethod() throws InterruptedException {
        logger.info("开始异步方法.");
        Thread.sleep(1000L);
        logger.info("异步方法结束.");
    }
}
```

7.1.3 创建类 SchedulingService

在包 com.bookcode.service 中创建类 SchedulingService,代码如例 7-3 所示。

【例 7-3】 创建类 SchedulingService 的代码示例。

```
package com.bookcode.service;
import org.slf4j.Logger;
import org.slf4j.LoggerFactory;
```

```java
import org.springframework.beans.factory.annotation.Autowired;
import org.springframework.scheduling.annotation.Scheduled;
import org.springframework.stereotype.Service;
//@Description: 定时服务
@Service
public class SchedulingService {
    private Logger logger = LoggerFactory.getLogger(this.getClass());
    private final SleuthService sleuthService;
    @Autowired
    public SchedulingService(SleuthService sleuthService) {
        this.sleuthService = sleuthService;
    }
    @Scheduled(fixedDelay = 30000)
    public void scheduledWork() throws InterruptedException {
        logger.info("从计划任务中开始一些工作.");
        sleuthService.asyncMethod();
        logger.info("从计划任务中结束工作.");
    }
}
```

7.1.4 创建类 ThreadConfig

在包 com.bookcode 中创建 config 子包，并在包 com.bookcode.config 中创建类 ThreadConfig，代码如例 7-4 所示。

【例 7-4】 创建类 ThreadConfig 的代码示例。

```java
package com.bookcode.config;
import java.util.concurrent.Executor;
import java.util.concurrent.Executors;
import org.springframework.beans.factory.BeanFactory;
import org.springframework.beans.factory.annotation.Autowired;
import org.springframework.cloud.sleuth.instrument.async.LazyTraceExecutor;
import org.springframework.context.annotation.Bean;
import org.springframework.context.annotation.Configuration;
import org.springframework.scheduling.annotation.AsyncConfigurerSupport;
import org.springframework.scheduling.annotation.EnableAsync;
import org.springframework.scheduling.annotation.EnableScheduling;
import org.springframework.scheduling.annotation.SchedulingConfigurer;
import org.springframework.scheduling.concurrent.ThreadPoolTaskExecutor;
import org.springframework.scheduling.config.ScheduledTaskRegistrar;
@Configuration
@EnableAsync
@EnableScheduling
public class ThreadConfig extends AsyncConfigurerSupport implements SchedulingConfigurer {
```

```java
    @Autowired
    private BeanFactory beanFactory;
    @Override
    public Executor getAsyncExecutor() {
        ThreadPoolTaskExecutor threadPoolTaskExecutor = new ThreadPoolTaskExecutor();
        threadPoolTaskExecutor.setCorePoolSize(7);
        threadPoolTaskExecutor.setMaxPoolSize(42);
        threadPoolTaskExecutor.setQueueCapacity(11);
        threadPoolTaskExecutor.setThreadNamePrefix("MyExecutor-");
        threadPoolTaskExecutor.initialize();
        return new LazyTraceExecutor(beanFactory, threadPoolTaskExecutor);
    }
    @Override
    public void configureTasks(ScheduledTaskRegistrar scheduledTaskRegistrar) {
        scheduledTaskRegistrar.setScheduler(schedulingExecutor());
    }
    @Bean(destroyMethod = "shutdown")
    public Executor schedulingExecutor() {
        return Executors.newScheduledThreadPool(1);
    }
}
```

7.1.5 创建类 HelloController

在包 com.bookcode 中创建 controller 子包,并在包 com.bookcode.controller 中创建类 HelloController,代码如例 7-5 所示。

【例 7-5】 创建类 HelloController 的代码示例。

```java
package com.bookcode.controller;
import com.bookcode.service.SleuthService;
import org.slf4j.Logger;
import org.slf4j.LoggerFactory;
import org.springframework.beans.factory.annotation.Autowired;
import org.springframework.web.bind.annotation.GetMapping;
import org.springframework.web.bind.annotation.RestController;
@RestController
public class HelloController {
    private Logger logger = LoggerFactory.getLogger(this.getClass());
    private final SleuthService sleuthService;
    @Autowired
    public HelloController(SleuthService sleuthService) {
        this.sleuthService = sleuthService;
    }
    @GetMapping("/same-span")
```

```
    public String helloSleuthSameSpan() throws InterruptedException {
        logger.info("Same Span");
        sleuthService.doSomeWorkSameSpan();
        return "成功完成一些 Span.";
    }
    @GetMapping("/async")
    public String helloSleuthAsync() throws InterruptedException {
        logger.info("在异步方法调用之前.");
        sleuthService.asyncMethod();
        logger.info("在异步方法调用之后.");
        return "成功完成异步方法.";
    }
}
```

7.1.6 修改配置文件 application.properties

修改配置文件 application.properties,修改后的代码如例 7-6 所示。

【例 7-6】 修改后的配置文件 application.properties 代码示例。

```
spring.application.name = sleuthexample
```

7.1.7 运行程序

运行程序后,控制台输出信息如例 7-7 所示,结果如图 7-1 所示。

【例 7-7】 控制台输出信息示例。

```
2018 - 11 - 21 07:39:01.028 INFO [sleuthexample,69fbe286c8e42dc7,69fbe286c8e42dc7,false]
13624 --- [pool - 1 - thread - 1] com.bookcode.service.SchedulingService   : 从计划任务中
开始一些工作.
2018 - 11 - 21 07:39:01.032 INFO [sleuthexample,69fbe286c8e42dc7,69fbe286c8e42dc7,false]
13624 --- [pool - 1 - thread - 1] com.bookcode.service.SchedulingService   : 从计划任务中
结束工作.
2018 - 11 - 21 07:39:01.035 INFO [sleuthexample,69fbe286c8e42dc7,7a89bbfd62891a58,false]
13624 --- [ MyExecutor - 1] com.bookcode.service.SleuthService   : 开始异步方法.
2018 - 11 - 21 07:39:01.058 INFO [sleuthexample,,,] 13624 --- [         main] o.s.b.w.
embedded.tomcat.TomcatWebServer : Tomcat started on port(s): 8080 (http) with context path ''
2018 - 11 - 21 07:39:01.062 INFO [sleuthexample,,,] 13624 --- [         main] com.bookcode.
DemoApplication       : Started DemoApplication in 12.126 seconds (JVM running for 15.076)
2018 - 11 - 21 07:39:02.035 INFO [sleuthexample,69fbe286c8e42dc7,7a89bbfd62891a58,false]
13624 --- [  MyExecutor - 1] com.bookcode.service.SleuthService   : 异步方法结束.
spring.application.name = sleuthexample
```

在浏览器中输入 localhost:8080/async,浏览器中的结果如图 7-2 所示,控制台中结果如图 7-3 所示。在浏览器中输入 localhost:8080/same-span,浏览器中的结果如图 7-4 所示,控制台中的结果如图 7-5 所示。

图 7-1　运行程序后控制台中的结果

图 7-2　在浏览器中输入 localhost:8080/async 后浏览器中的结果

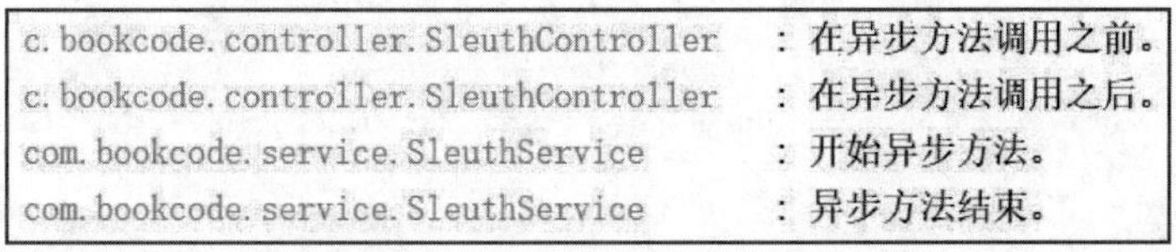

c.bookcode.controller.SleuthController	：在异步方法调用之前。
c.bookcode.controller.SleuthController	：在异步方法调用之后。
com.bookcode.service.SleuthService	：开始异步方法。
com.bookcode.service.SleuthService	：异步方法结束。

图 7-3　在浏览器中输入 localhost:8080/async 后控制台中的结果

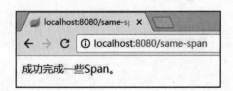

图 7-4　在浏览器中输入 localhost:8080/same-span 后浏览器中的结果

com.bookcode.service.SchedulingService	：从计划任务中开始一些工作。
com.bookcode.service.SchedulingService	：从计划任务中结束工作。
com.bookcode.service.SleuthService	：开始异步方法。
c.bookcode.controller.SleuthController	：Same Span
com.bookcode.service.SleuthService	：异步方法结束。
com.bookcode.service.SleuthService	：模拟在做一些事情。

图 7-5　在浏览器中输入 localhost:8080/same-span 后控制台中的结果

7.2　Spring Cloud Zipkin 的应用

视频讲解

Zipkin 是一个开源项目，它提供了在分布式环境下发送、接收、存储和可视化跟踪的机制。Zipkin 服务将存储对服务的所有操作步骤，每一步操作都会发送到该服务器，用于跟踪识别。本节介绍 Spring Cloud Zipkin 的应用。

7.2.1　创建项目 zipkinexample

用 IDEA 创建完项目 zipkinexample 之后，确保 pom.xml 文件中关于版本和依赖的代码如例 7-7 所示。

【例 7-7】 pom.xml 文件中关于版本和依赖的代码示例。

```xml
<parent>
    <groupId>org.springframework.boot</groupId>
    <artifactId>spring-boot-starter-parent</artifactId>
    <!-- 注意版本信息,升级到 Spring Boot 2.0 后会出错 -->
    <version>1.5.16.RELEASE</version>
    <relativePath/> <!-- lookup parent from repository -->
</parent>
<properties>
    <project.build.sourceEncoding>UTF-8</project.build.sourceEncoding>
    <project.reporting.outputEncoding>UTF-8</project.reporting.outputEncoding>
    <java.version>1.8</java.version>
    <!-- 注意版本信息,升级到 Spring Boot 2.0 配套的 Finchley.RELEASE 会出错 -->
    <spring-cloud.version>Edgware.RELEASE</spring-cloud.version>
</properties>
<dependencies>
    <dependency>
        <groupId>io.zipkin.java</groupId>
        <artifactId>zipkin-autoconfigure-ui</artifactId>
    </dependency>
```

```xml
<dependency>
    <groupId>io.zipkin.java</groupId>
    <artifactId>zipkin-server</artifactId>
    <version>2.4.0</version>
</dependency>
<dependency>
    <groupId>org.springframework.boot</groupId>
    <artifactId>spring-boot-starter-web</artifactId>
</dependency>
<dependency>
    <groupId>org.springframework.boot</groupId>
    <artifactId>spring-boot-starter-test</artifactId>
    <scope>test</scope>
</dependency>
</dependencies>
```

在目录 src/main/resources 下，创建配置文件 application.yml，代码如例 7-8 所示。

【例 7-8】 创建配置文件 application.yml 的代码示例。

```yaml
spring:
    application:
        name: zipkinexample
server:
    port: 9411
```

修改入口类，修改后的代码如例 7-9 所示。

【例 7-9】 修改后的入口类的代码示例。

```java
package com.bookcode;
import org.springframework.boot.SpringApplication;
import org.springframework.boot.autoconfigure.SpringBootApplication;
import zipkin.server.*;
@SpringBootApplication
@EnableZipkinServer
public class DemoApplication {
    public static void main(String[] args) {
        SpringApplication.run(DemoApplication.class, args);
    }
}
```

7.2.2 创建项目 zipkinclient1

用 IDEA 创建完项目 zipkinclient1 之后，确保 pom.xml 文件中关于版本和依赖的代码如例 7-10 所示。

【例 7-10】 pom.xml 文件中关于版本和依赖的代码示例。

```xml
<parent>
    <groupId>org.springframework.boot</groupId>
    <artifactId>spring-boot-starter-parent</artifactId>
    <version>1.5.16.RELEASE</version>
    <relativePath/> <!-- lookup parent from repository -->
</parent>
<properties>
    <project.build.sourceEncoding>UTF-8</project.build.sourceEncoding>
    <project.reporting.outputEncoding>UTF-8</project.reporting.outputEncoding>
    <java.version>1.8</java.version>
    <spring-cloud.version>Edgware.RELEASE</spring-cloud.version>
</properties>
<dependencies>
    <dependency>
        <groupId>org.springframework.boot</groupId>
        <artifactId>spring-boot-starter-web</artifactId>
    </dependency>
    <dependency>
        <groupId>org.springframework.cloud</groupId>
        <artifactId>spring-cloud-starter-zipkin</artifactId>
    </dependency>
    <dependency>
        <groupId>org.springframework.boot</groupId>
        <artifactId>spring-boot-starter-test</artifactId>
        <scope>test</scope>
    </dependency>
</dependencies>
```

修改配置文件 application.properties，修改后的代码如例 7-11 所示。

【例 7-11】 修改后的配置文件 application.properties 的代码示例。

```
spring.application.name = zipkinclient1-user3
server.port = 9433
spring.zipkin.locator.discovery.enabled=true
spring.zipkin.baseUrl=http://localhost:9411/
spring.sleuth.sampler.percentage = 1.0
spring.sleuth.web.skipPattern = (^cleanup.*)
```

在包 com.bookcode 中创建 controller 子包，并在包 com.bookcode.controller 中创建类 HelloController，代码如例 7-12 所示。

【例 7-12】 创建类 HelloController 的代码示例。

```
package com.bookcode.controller;
import org.springframework.beans.factory.annotation.Value;
import org.springframework.web.bind.annotation.GetMapping;
import org.springframework.web.bind.annotation.RestController;
@RestController
public class HelloController {
    @Value("${server.port}")
    private String port;
    @GetMapping("/hello")
    public String hello() {
        return "Hello, I am Zipkin Client " + "at " + port + ".";
    }
}
```

7.2.3 创建项目 zipkinuser1

用 IDEA 创建完项目 zipkinuser1 之后,确保 pom.xml 文件中关于版本和依赖的代码如例 7-10 所示。

修改配置文件 application.properties,修改后的代码如例 7-13 所示。

【例 7-13】 修改后的配置文件 application.properties 的代码示例。

```
spring.application.name = zipkinuser1-role3
server.port = 9414
spring.zipkin.locator.discovery.enabled = true
spring.zipkin.baseUrl = http://localhost:9411/
spring.sleuth.sampler.percentage = 1.0
spring.sleuth.web.skipPattern = (^cleanup.*)
```

在包 com.bookcode 中创建 controller 子包,并在包 com.bookcode.controller 中创建类 HiController,代码如例 7-14 所示。

【例 7-14】 创建类 HiController 的代码示例。

```
package com.bookcode.controller;
import org.springframework.beans.factory.annotation.Value;
import org.springframework.web.bind.annotation.GetMapping;
import org.springframework.web.bind.annotation.RestController;
@RestController
public class HiController {
    @Value("${server.port}")
    private String port;
```

```
@GetMapping("/hi")
public String hi() {
    return "Hi, I am Zipkin User " + "at " + port + " .";
}
}
```

7.2.4 运行程序

依次运行程序 zipkinclient1、zipkinuser1 和 zipkinexample 后,在浏览器中输入 localhost:9433/hello,结果如图 7-6 所示。在浏览器中输入 localhost:9414/hi,结果如图 7-7 所示。在浏览器中输入 localhost:9411,单击 Find Traces 按钮,结果如图 7-8 所示。针对项目 zipkinuser1,单击 JSON 按钮,结果如图 7-9 所示。

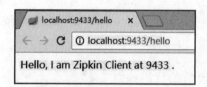

图 7-6 在浏览器中输入 localhost:9433/hello 的结果

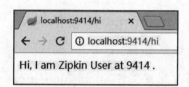

图 7-7 在浏览器中输入 localhost:9414/hi 的结果

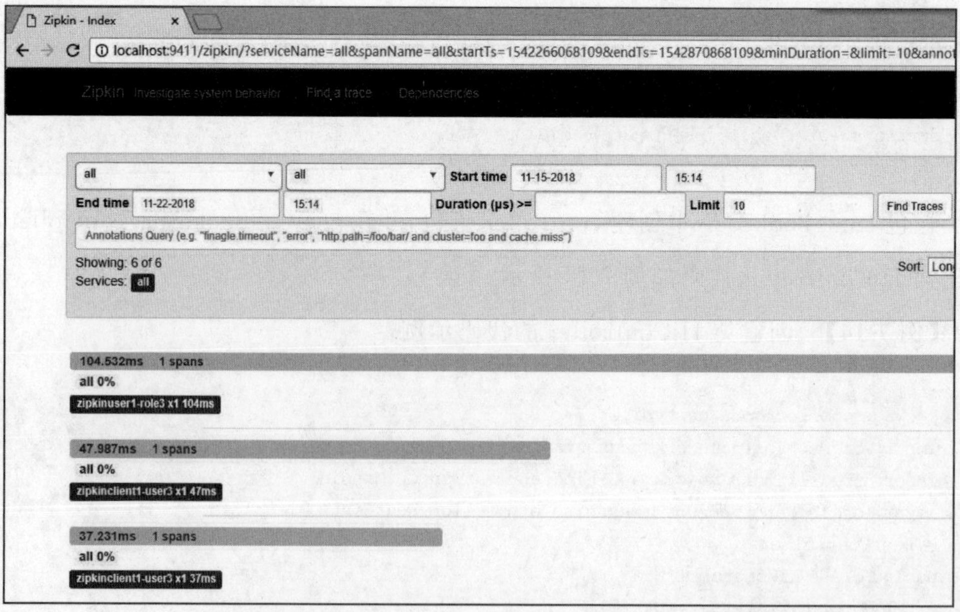

图 7-8 在浏览器中输入 localhost:9411 后单击 Find Traces 按钮的结果

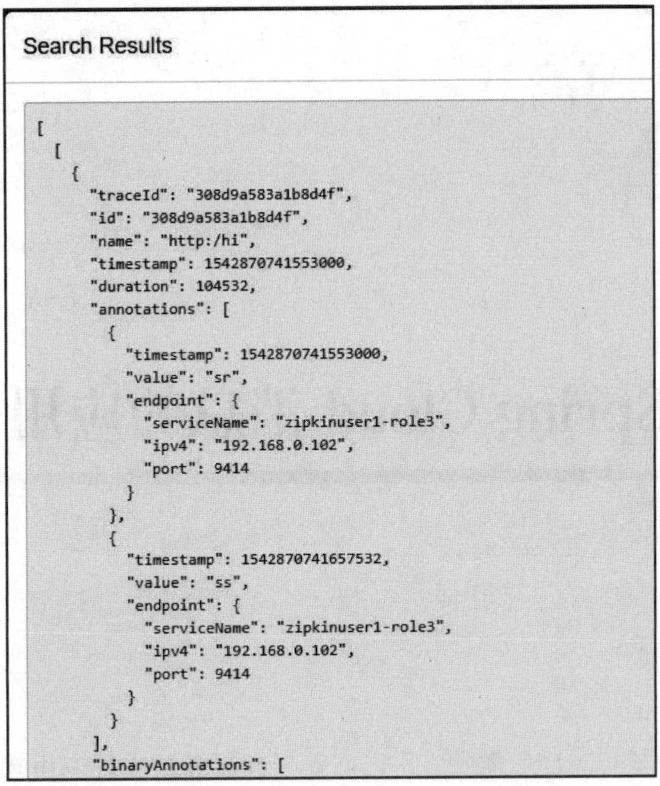

图 7-9　针对项目 zipkinuser1 单击 JSON 按钮的结果

习题 7

实验题

1. 请实现 Spring Cloud Sleuth 的应用开发。
2. 请实现 Spring Cloud Zipkin 的应用开发。

第 8 章

Spring Cloud 消息的应用

企业应用集成(EAI)是集成应用之间数据和服务的一种应用技术。有文件传输、共享数据库、远程过程调用和消息等 4 种集成风格。消息风格是指两个系统连接到一个公用的消息系统，互相交换数据，并利用消息调用行为。

消息总线扮演着一种消息路由的角色，拥有一套完备的路由机制来决定消息传输方向。发送端只需要向消息总线发出消息而不用管消息被如何转发。本章介绍 Spring Cloud Bus 和 Spring Cloud Stream 的应用。

8.1 Spring Cloud Bus 的应用

视频讲解

Spring Cloud Bus 通过轻量消息代理连接各个分布的节点。管理和传播所有分布式项目中的消息，其本质是利用了消息列队 MQ 的广播机制在分布式系统中传播消息，目前常用的有 Kafka 和 RabbitMQ 等。本节介绍 Spring Cloud Bus 和 RabbitMQ 的综合应用。

8.1.1 Spring Cloud Config Server 的应用

用 IDEA 创建完项目 configserverwitheureka 之后，确保在文件 pom.xml 的

<dependencies>和</dependencies>之间添加了 Config Server 和 Eureka Client 依赖,代码如例 8-1 所示。

【例 8-1】 添加 Config Server 和 Eureka Client 依赖的代码示例。

```
<dependency>
        <groupId>org.springframework.cloud</groupId>
        <artifactId>spring-cloud-config-server</artifactId>
</dependency>
<dependency>
        <groupId>org.springframework.cloud</groupId>
        <artifactId>spring-cloud-starter-netflix-eureka-client</artifactId>
</dependency>
```

修改配置文件 application.properties,修改后的代码如例 8-2 所示。

【例 8-2】 修改后的配置文件 application.properties 代码示例。

```
spring.application.name = configserverwitheureka
server.port = 8768
spring.cloud.config.server.git.uri = https://github.com/woodsheng/MicroserviceWithSpringCloud.git
spring.cloud.config.server.git.username =
spring.cloud.config.server.git.password =
eureka.client.serviceUrl.defaultZone = http://localhost:8761/eureka/
management.endpoints.web.exposure.include = bus-refresh
```

修改入口类,修改后的代码如例 8-3 所示。

【例 8-3】 修改后的入口类的代码示例。

```java
package com.bookcode;
import org.springframework.boot.SpringApplication;
import org.springframework.boot.autoconfigure.SpringBootApplication;
import org.springframework.cloud.client.discovery.EnableDiscoveryClient;
import org.springframework.cloud.config.server.EnableConfigServer;
@SpringBootApplication
@EnableConfigServer
@EnableDiscoveryClient
public class DemoApplication {
    public static void main(String[] args) {
        SpringApplication.run(DemoApplication.class, args);
    }
}
```

8.1.2 Spring Cloud Bus 的应用实现

用 IDEA 创建完项目 busexample 之后，确保在文件 pom.xml 的<dependencies>和</dependencies>之间添加了 Eureka Client、Amqp、Config、Web 等依赖，代码如例 8-4 所示。

【例 8-4】 添加 Eureka Client、Amqp、Config、Web 等依赖的代码示例。

```xml
<dependency>
    <groupId>org.springframework.cloud</groupId>
    <artifactId>spring-cloud-starter-netflix-eureka-client</artifactId>
</dependency>
<dependency>
    <groupId>org.springframework.cloud</groupId>
    <artifactId>spring-cloud-starter-bus-amqp</artifactId>
</dependency>
<dependency>
    <groupId>org.springframework.cloud</groupId>
    <artifactId>spring-cloud-starter-config</artifactId>
</dependency>
<dependency>
    <groupId>org.springframework.boot</groupId>
    <artifactId>spring-boot-starter-web</artifactId>
</dependency>
<dependency>
    <groupId>org.springframework.boot</groupId>
    <artifactId>spring-boot-starter-actuator</artifactId>
</dependency>
```

在目录 src/main/resources 下，创建并修改配置文件 bootstrap.properties，修改后的代码如例 8-5 所示。

【例 8-5】 修改后的配置文件 bootstrap.properties 代码示例。

```
spring.application.name = config-client
spring.cloud.config.label = master
spring.cloud.config.profile = dev
eureka.client.serviceUrl.defaultZone = http://localhost:8761/eureka/
spring.cloud.config.discovery.enabled = true
spring.cloud.config.discovery.serviceId = configserverwitheureka
server.port = 8882              #或者使用端口 8881
spring.rabbitmq.host = localhost
spring.rabbitmq.port = 5672
spring.rabbitmq.username = guest
spring.rabbitmq.password = guest
spring.cloud.bus.enabled = true
spring.cloud.bus.trace.enabled = true
management.endpoints.web.exposure.include = bus-refresh
```

修改入口类，修改后的代码如例 8-6 所示。

【例 8-6】 修改后的入口类的代码示例。

```
package com.bookcode;
import org.springframework.beans.factory.annotation.Value;
import org.springframework.boot.SpringApplication;
import org.springframework.boot.autoconfigure.SpringBootApplication;
import org.springframework.cloud.client.discovery.EnableDiscoveryClient;
import org.springframework.cloud.context.config.annotation.RefreshScope;
import org.springframework.web.bind.annotation.RequestMapping;
import org.springframework.web.bind.annotation.RestController;
@SpringBootApplication
@EnableDiscoveryClient
@RestController
@RefreshScope
public class DemoApplication {
    @Value("${foo}")
    String foo;
    @RequestMapping(value = "/hi")
    public String hi(){
        return foo;
    }
    public static void main(String[] args) {
        SpringApplication.run(DemoApplication.class, args);
    }
}
```

8.1.3 运行程序

依次启动 3.2 节中 eureka-server 程序和本节的 configserverwitheureka 程序，启动本节 busexample 程序的两个实例，端口分别为 8881、8882（其他内容不变，但在例 8-5 中设置不同的端口）。运行程序后，在浏览器中输入 localhost:8881/hi，结果如图 8-1 所示。在浏览器中输入 localhost:8882/hi，结果如图 8-2 所示。

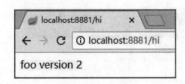

图 8-1　在浏览器中输入 localhost: 8881/hi 的结果

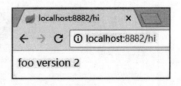

图 8-2　在浏览器中输入 localhost: 8882/hi 的结果

执行如例 8-7 所示的命令启动 RabbitMQ 服务。

【例 8-7】 启动 RabbitMQ 服务的命令示例。

```
rabbitmq-server
```

在浏览器中输入 localhost:15672/#/connections，结果如图 8-3 所示。将 GitHub 代码仓库(https://github.com/)中 foo 值改为 Update foo to version 4，即改变配置文件 foo 的值。传统做法需要重启服务才能达到配置文件的更新。本例中，只需要发送 POST 请求 http://localhost:8881/actuator/bus-refresh(本例是在工具 Postman 中发送 POST 请求)，如图 8-4 所示；busexample 会重新读取配置文件。在浏览器中输入 localhost:8881/hi，结果如图 8-5 所示。在浏览器中输入 localhost:8882/hi，结果如图 8-6 所示。

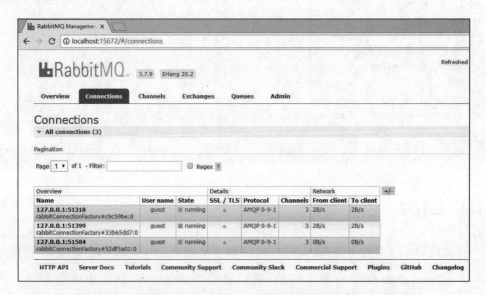

图 8-3 在浏览器中输入 localhost:15672/#/connections 的结果

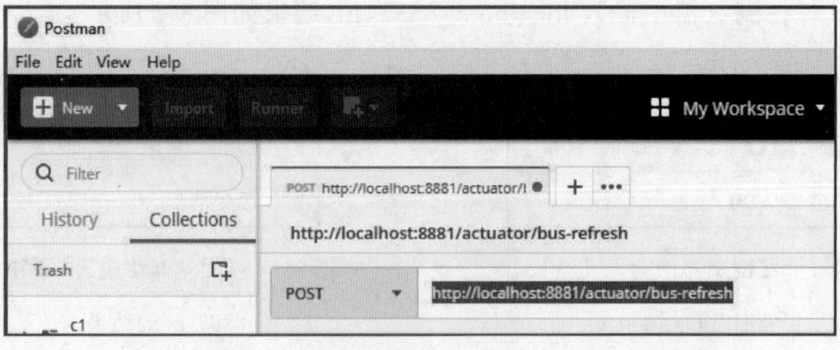

图 8-4 在工具 Postman 中发送 POST 请求(http://localhost:8881/actuator/bus-refresh)的结果

图 8-5　在浏览器中输入 localhost：8881/hi 的结果

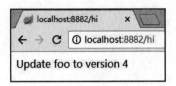

图 8-6　在浏览器中输入 localhost：8882/hi 的结果

8.2　Spring Cloud Stream 的应用

视频讲解

Spring Cloud Stream 是一个用于为微服务应用构建消息驱动能力的框架。它通过使用 Spring Integration 来连接消息代理中间件，以实现消息事件驱动。Spring Cloud Stream 为一些消息中间件产品提供了个性化的自动化配置实现，引用了发布-订阅的通信模型、订阅相同目标（即发布者）的消费者为一个消费组、消息分区这 3 个核心概念。目前仅支持 RabbitMQ、Kafka。

8.2.1　创建项目并添加依赖

用 IDEA 创建完项目 streamexample 之后，确保在文件 pom.xml 的 < dependencies > 和 </dependencies > 之间添加了 Stream Rabbit 依赖，代码如例 8-8 所示。

【例 8-8】　添加 Stream Rabbit 依赖的代码示例。

```
< dependency >
        < groupId > org.springframework.cloud </groupId >
        < artifactId > spring-cloud-starter-stream-rabbit </artifactId >
        < version > 2.0.0.RELEASE </version >
</dependency >
```

8.2.2　创建接口 Sink

创建接口 Sink，代码如例 8-9 所示。

【例 8-9】　创建接口 Sink 的代码示例。

```
package com.bookcode;
import org.springframework.cloud.stream.annotation.Input;
```

```
import org.springframework.messaging.SubscribableChannel;
public interface Sink {
    String INPUT = "input";
    @Input("input")
    SubscribableChannel input();
}
```

8.2.3 创建类 SinkReceiver

创建类 SinkReceiver，代码如例 8-10 所示。

【例 8-10】 创建类 SinkReceiver 的代码示例。

```
package com.bookcode;
import org.slf4j.Logger;
import org.slf4j.LoggerFactory;
import org.springframework.cloud.stream.annotation.EnableBinding;
import org.springframework.cloud.stream.annotation.StreamListener;
@EnableBinding(value = {Sink.class})          //用来指定定义@Input 或者@Output 注解的接口
//实现对消息通道的绑定.Sink 接口是默认输入消息通道绑定接口
public class SinkReceiver {
    private static Logger logger = LoggerFactory.getLogger(SinkReceiver.class);
    @StreamListener(Sink.INPUT) //将被修饰的方法注册为消息中间件上数据流的事件监听器
    public void receive(String payload) {
            logger.info("SinkReceiver---接收到的信息: " + payload);
    }
}
```

8.2.4 创建配置文件 application.yml

在目录 src/main/resources 下，创建配置文件 application.yml，代码如例 8-11 所示。

【例 8-11】 创建配置文件 application.yml 的代码示例。

```
logging:
    file: log\outputinfo.log
```

8.2.5 运行程序

执行如例 8-7 所示的命令，启动 RabbitMQ 服务。

运行程序后，在浏览器中输入 localhost:15672/#/Queues，并在 Payload 文本框

中输入"您好，这是 RabbitMQ 发布的信息！"，单击 Publish message 按钮，弹出提示信息框（内容为"Message published"），结果如图 8-7 所示。完成上述操作后，控制台中的相关输出如图 8-8 所示。与此同时，根据配置文件的设置在项目中自动创建了一个子目录（\log），并在子目录中生成了一个日志文件（outputinfo.log）；日志文件内容中的相关输出结果如图 8-9 所示。

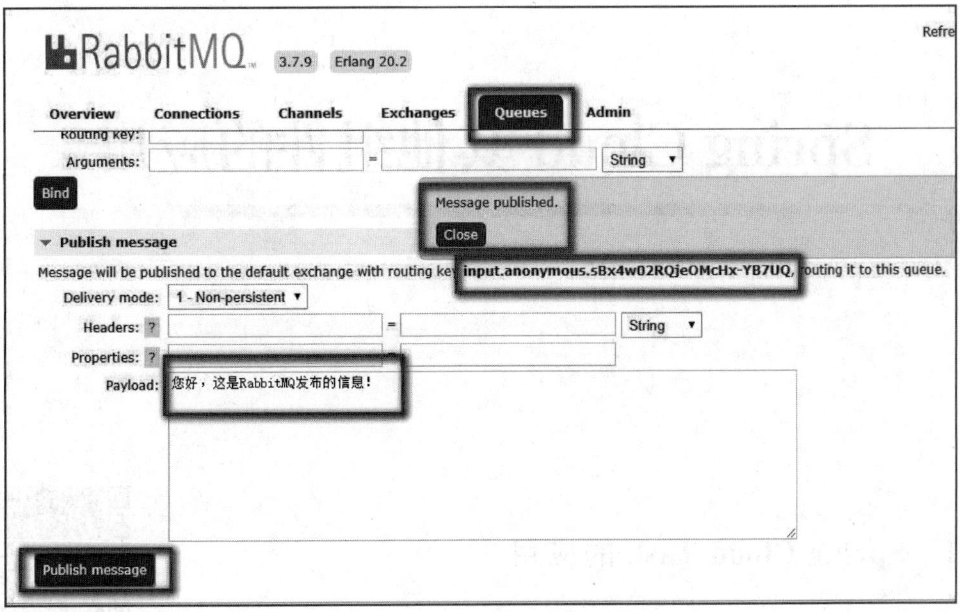

图 8-7　在浏览器中输入 localhost:15672/♯/Queues 并进行相关操作的结果

图 8-8　完成相关操作后控制台中的相关输出结果

图 8-9　完成相关操作后日志文件内容中的相关输出结果

习题 8

实验题

1. 请实现 Spring Cloud Bus 的应用开发。
2. 请实现 Spring Cloud Stream 的应用开发。

第 9 章

Spring Cloud 其他组件的应用

9.1 Spring Cloud Task 的应用

视频讲解

本章介绍 Spring Cloud Task、Spring Cloud Function 和 Cloud Foundry 的应用。

9.1.1 创建项目并添加依赖

用 IDEA 创建完项目 taskexample 之后,确保在文件 pom.xml 的< dependencies >和</ dependencies >之间添加了 Task、Web 依赖,代码如例 9-1 所示。

【例 9-1】 添加 Task、Web 依赖的代码示例。

```
<dependency>
        <groupId>org.springframework.cloud</groupId>
        <artifactId>spring-cloud-starter-task</artifactId>
</dependency>
<dependency>
        <groupId>org.springframework.boot</groupId>
        <artifactId>spring-boot-starter-web</artifactId>
</dependency>
```

9.1.2 创建类 ScheduledTask

创建类 ScheduledTask，代码如例 9-2 所示。

【例 9-2】 创建类 ScheduledTask 的代码示例。

```java
package com.bookcode;
import org.springframework.scheduling.annotation.Scheduled;
import org.springframework.stereotype.Component;
import java.text.SimpleDateFormat;
import java.util.Date;
@Component
public class ScheduledTask {
    SimpleDateFormat sdf = new SimpleDateFormat("yyyy年MM月dd日 HH时mm分ss秒");
    @Scheduled(fixedRate = 10000)
    public void testFixRate() {
        System.out.println("每隔10秒执行一次：" + sdf.format(new Date()));
    }
}
```

9.1.3 创建类 HelloController

在包 com.bookcode 中创建 controller 子包，并在包 com.bookcode.controller 中创建类 HelloController，代码如例 9-3 所示。

【例 9-3】 创建类 HelloController 的代码示例。

```java
package com.bookcode.controller;
import org.springframework.web.bind.annotation.GetMapping;
import org.springframework.web.bind.annotation.RestController;
@RestController
public class HelloController {
    @GetMapping("/hello")
    public String hello() {
        return "Hello, Task.";
    }
}
```

9.1.4 创建配置文件 application.yml

在目录 src/main/resources 下，创建配置文件 application.yml，代码如例 9-4 所示。

【例 9-4】 创建配置文件 application.yml 的代码示例。

```yaml
spring:
  application:
    name: HelloWorld
logging:
  level:
    org:
      springframework:
        cloud:
          task: DEBUG
```

9.1.5 修改入口类

修改入口类,修改后的代码如例 9-5 所示。

【例 9-5】 修改后的入口类代码示例。

```java
package com.bookcode;
import org.springframework.boot.SpringApplication;
import org.springframework.boot.autoconfigure.SpringBootApplication;
import org.springframework.cloud.task.configuration.EnableTask;
import org.springframework.scheduling.annotation.EnableScheduling;
@SpringBootApplication
@EnableTask
@EnableScheduling
public class DemoApplication {
    public static void main(String[] args) {
        SpringApplication.run(DemoApplication.class, args);
    }
}
```

9.1.6 运行程序

运行程序,控制台的输出如图 9-1 所示。在浏览器中输入 localhost:8080/hello,结果如图 9-2 所示。

```
每隔10秒执行一次:2018年12月04日 18时45分01秒
每隔10秒执行一次:2018年12月04日 18时45分11秒
每隔10秒执行一次:2018年12月04日 18时45分21秒
```

图 9-1 控制台中的输出结果

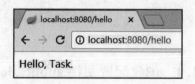

图 9-2 在浏览器中输入 localhost:8080/hello 后的结果

9.2 Spring Cloud Function 的应用

视频讲解

9.2.1 创建项目并添加依赖

用 IDEA 创建完项目 functionexample 之后,确保在文件 pom.xml 的< dependencies >和</dependencies >之间添加了 Function Web 依赖,代码如例 9-6 所示。

【例 9-6】 添加 Function Web 依赖的代码示例。

```
<dependency>
        <groupId>org.springframework.cloud</groupId>
        <artifactId>spring-cloud-starter-function-web</artifactId>
</dependency>
```

9.2.2 创建类 Greeter

创建类 Greeter,代码如例 9-7 所示。

【例 9-7】 创建类 Greeter 的代码示例。

```
package com.bookcode;
import org.springframework.stereotype.Component;
import java.util.function.Function;
@Component
public class Greeter implements Function<String, String> {
    public String apply(String name) {
        return "Hello " + name;
    }
}
```

9.2.3 创建类 HelloController

在包 com.bookcode 中创建 controller 子包,并在包 com.bookcode.controller 中创建类 HelloController,代码如例 9-8 所示。

【例 9-8】 创建类 HelloController 的代码示例。

```
package com.bookcode.controller;
import org.springframework.beans.factory.annotation.Autowired;
```

```
import org.springframework.web.bind.annotation.GetMapping;
import org.springframework.web.bind.annotation.PathVariable;
import org.springframework.web.bind.annotation.RestController;
@RestController
public class HelloController {
    @Autowired
    private Greeter greeter;
    @GetMapping("/hello/{name}")
    public String hello(@PathVariable String name) {
        return greeter.apply(name);
    }
}
```

9.2.4 运行程序

运行程序,在浏览器中输入 localhost:8080/hello/张三,结果如图 9-3 所示。

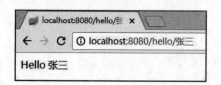

图 9-3 在浏览器中输入 localhost:8080/hello/张三后的结果

9.3 Cloud Foundry 的应用

视频讲解

9.3.1 Cloud Foundry 简介

Cloud Foundry 是一个开源 PaaS 云平台,它支持多种框架、语言、运行时环境、云平台及应用服务,使开发人员能够在几秒钟内进行应用程序的部署和扩展,而无须担心任何基础架构的问题。部署至 Cloud Foundry 的应用程序可通过 Open Service Broker API 访问外部资源。

在平台中,数据库、消息系统、文件系统等所有外部依赖项都被视为服务。当应用程序被推送到 Cloud Foundry 时,也可以指定它所需的服务。

9.3.2 利用 Cloud Foundry 平台部署 Spring Boot 应用

部署是开发的最后一步,也是较难的一步。一般云服务器都提供了很好的服务,如阿里云等。利用 Cloud Foundry 平台的好处是使用云服务、部署简单，而且有 2GB 的免费内存空间。

以 Windows 系统为例，利用 Cloud Foundry 平台部署 Spring Boot 应用，步骤如下所述。

（1）安装 Cloud Foundry 平台的 Windows installer。

（2）注册 Pivotal 账号。

（3）部署 Spring Boot 应用。选择自己要部署的 Spring Boot 项目，生成该项目的 jar 包。示例的 Spring Boot 项目为 Josepus（https://github.com/percent4/josephus）。在 Windows 命令处理程序 CMD 中切换到该 jar 包所在文件夹，然后输入如例 9-9 的命令来上传服务，结果如图 9-4 所示。

【例 9-9】上传服务的命令示例。

```
cf push josephus – p josephus – 0.1.0.jar
```

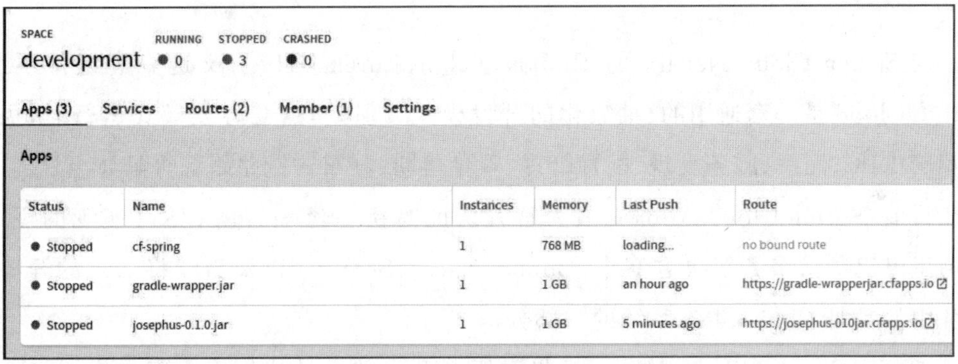

图 9-4　上传服务后的结果

习题 9

实验题

1. 请实现 Spring Cloud Task 的应用。
2. 请实现 Spring Cloud Function 的应用。
3. 请实现 Cloud Foundry 的应用。

第10章

Spring Cloud Alibaba 的应用

与 Spring Cloud Netflix 的 Ribbon、Feign、Eureka、Hystrix 这一套组件不同，Spring Cloud 是一套通用的、抽象化的开发模式。但是，这套开发模式运行时，仍需依赖于 RPC、网关、服务发现、配置管理、限流熔断、分布式链路跟踪等具体组件。

目前，Spring Cloud Alibaba 成为官方认证的新一套 Spring Cloud 实现的规范。在阿里巴巴的微服务解决方案中，Dubbo、Nacos、Sentinel 以及后续将加入的开源微服务组件都是 Dubbo 生态系统的一部分。

Spring Cloud Alibaba 致力于提供微服务开发的一站式解决方案。依托 Spring Cloud Alibaba，只需要添加一些注解和少量配置，就可以将 Spring Cloud 应用接入阿里应用，通过阿里中间件迅速搭建分布式应用系统。本节介绍 Spring Cloud Alibaba 的应用。

10.1 Spring Cloud Alibaba 简介

10.1.1 Spring Cloud Alibaba 主要功能

Spring Cloud Alibaba 采用阿里中间件作为原料，实现了 Spring Cloud 规范。

Spring Cloud Alibaba 包括服务限流降级、服务注册与发现、分布式配置管理、阿里云对象存储等功能。

其中，服务限流降级功能默认支持 Servlet、RestTemplate、Dubbo 和 RocketMQ 限流降级功能的接入，可以在运行时通过控制台实时修改限流降级规则，还支持查看限流降级度量监控。服务注册与发现功能提供了适配 Spring Cloud 的服务注册与发现标准，默认集成了对 Ribbon 的支持。分布式配置管理功能支持外部化配置，配置更改时自动刷新。对象存储服务（Object Storage Service，OSS 或简称为 OBS）功能提供了海量、安全、低成本、高可靠的云存储服务，支持在任何应用、任何时间、任何地点存储以及支持访问任意类型的数据。

10.1.2　Spring Cloud Alibaba 组件

Spring Cloud Alibaba 包含的组件内容，主要选取自阿里巴巴开源中间件和阿里云商业化产品，但也不限定于这些产品。目前发布的版本包含了 Nacos、Sentinel、OSS、ACM（Application Configuration Management）和 ANS（Application Naming Service）。随着阿里开源的力度不断加大，后续将会有更多的组件加入到 Spring Cloud 的实现中来。

Sentinel 是轻量级高可用流量控制组件。Sentinel 把流量作为切入点，从流量控制、熔断降级、系统负载保护等多个维度保护服务的稳定性。Sentinel 与 Hystrix 具有一些共同特性，但底层的实现方式不一致。Sentinel 具有轻量级、高性能、多样化的流量控制策略以及系统负载保护等特性。

Nacos 是一个更易于构建云原生应用的动态服务发现、配置管理和服务管理平台。经历过多次"双 11"检验，支持超大规模集群，其稳定性和性能上值得用户信赖。Nacos Discovery 实现了 Spring Cloud 服务发现的标准，也很好地适配了 Ribbon。相比之下，Nacos Discovery 支持实时推送，达到秒级服务发现；提供多层容灾机制，尽量保证服务发现中心宕机不影响应用调用；依赖组件更少，使用更加简单，功能更加强大。例如，Nacos Discovery 无须 Spring Cloud Bus、MQ，即可实现配置的实时推送，而且推送状态可查；支持数据回滚、多种数据格式及中文模式。

阿里巴巴的消息中间件 RocketMQ 是基于 Java 的高性能、高吞吐量的分布式消息和流计算平台。

AliCloud SchedulerX 是一套高可靠的分布式任务调度系统，支持海量任务秒级别调度,支持 Java、脚本、HTTP 等多种任务类型,接入简单,使用方便。

阿里巴巴的 OSS 提供海量、安全、低成本、高可靠的云存储服务,利用它可以在任何应用、任何时间、任何地点存储和访问任意类型的数据。它具有与平台无关的 RESTful API 接口,能够提供 99.999999999% 的数据可靠性和 99.99% 的服务可用性。

阿里巴巴的简单日志服务(Simple Log Service,SLS)是针对日志类数据的一站式服务,在阿里巴巴集团经历大量大数据场景锤炼而成。应用 SLS 实现无须开发就能快捷完成日志数据采集、消费、投递以及查询分析等功能,从而提升运维、运营效率,形成数据技术(Data Technology,DT)时代海量日志处理能力。

10.2 Nacos Config 的应用

视频讲解

10.2.1 创建项目并添加依赖

用 IDEA 创建完项目 nacosconfigexample 之后,确保在文件 pom.xml 的 <dependencies>和</dependencies>之间添加了 Nacos Config 和 Web 依赖,代码如例 10-1 所示。

【例 10-1】 添加 Nacos Config 和 Web 依赖的代码示例。

```
<dependency>
        <groupId>org.springframework.cloud</groupId>
        <artifactId>spring-cloud-starter-alibaba-nacos-config</artifactId>
        <version>0.2.0.RELEASE</version>
</dependency>
<dependency>
        <groupId>org.springframework.boot</groupId>
        <artifactId>spring-boot-starter-web</artifactId>
</dependency>
```

10.2.2 创建类 ConfigController

在包 com.bookcode 中创建 controller 子包,并在包 com.bookcode.controller 中创建类 ConfigController,代码如例 10-2 所示。

【例10-2】 创建类 ConfigController 的代码示例。

```
package com.bookcode.controller;
import org.springframework.beans.factory.annotation.Value;
import org.springframework.cloud.context.config.annotation.RefreshScope;
import org.springframework.web.bind.annotation.RequestMapping;
import org.springframework.web.bind.annotation.RestController;
@RestController
@RequestMapping("/config")
@RefreshScope
public class ConfigController {
    @Value("${useLocalCache:false}")
    private boolean useLocalCache;
    @RequestMapping("/get")
    public boolean get() {
        return useLocalCache;
    }
}
```

10.2.3　创建并修改配置文件 bootstrap.properties

在目录 src/main/resources 下，创建并修改配置文件 bootstrap.properties，修改后的代码如例 10-3 所示。

【例10-3】 修改后的配置文件 bootstrap.properties 的代码示例。

```
spring.cloud.nacos.config.server-addr=127.0.0.1:8848
spring.application.name=example
```

10.2.4　运行程序

下载、解压 Nacos Server 文件，打开 Windows 命令处理程序 CMD，执行如例 10-4 所示的命令启动 Nacos Server。

【例10-4】 启动 Nacos Server 的命令示例。

```
startup
```

在工具 Postman 中，发送 POST 请求，输入 http://127.0.0.1:8848/nacos/v1/cs/configs?dataId=example.properties&group=DEFAULT_GROUP&content=

useLocalCache=true，单击 Send 按钮，结果如图 10-1 所示。在浏览器中输入 localhost:8080/config/get，结果如图 10-2 所示。

在工具 Postman 中，再次发送 POST 请求，输入 http://127.0.0.1:8848/nacos/v1/cs/configs? dataId = example.properties&group = DEFAULT_GROUP&content = useLocalCache=false，单击 Send 按钮，结果如图 10-3 所示。再次在浏览器中输入 localhost:8080/config/get，结果如图 10-4 所示。

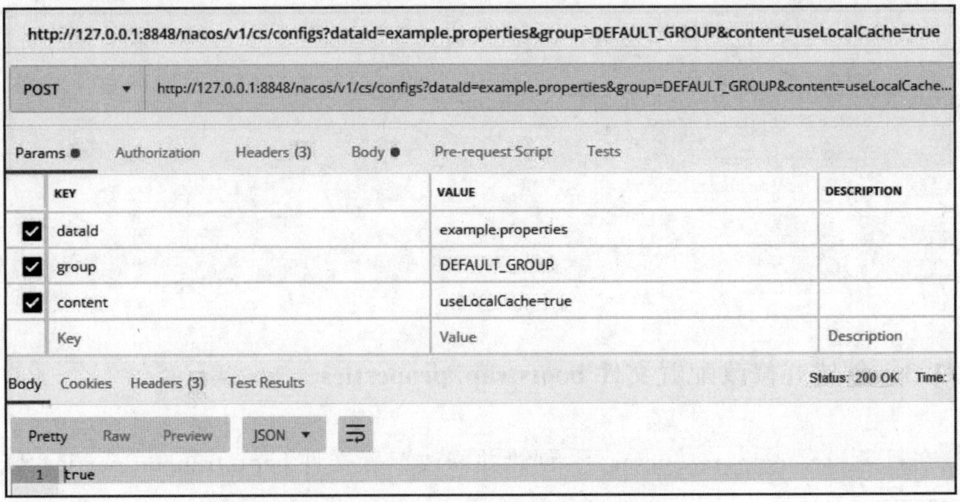

图 10-1 在 Postman 中输入 POST 请求的结果

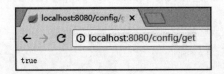

图 10-2 在浏览器中输入 localhost:8080/config/get 的结果

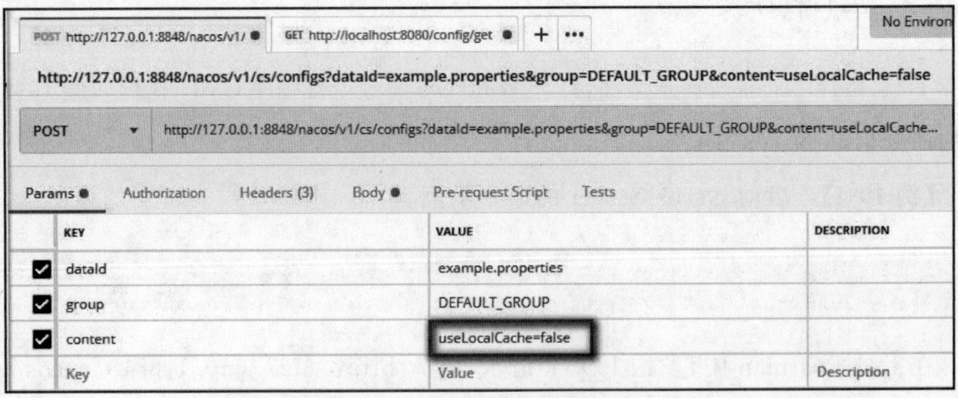

图 10-3 再次在 Postman 中输入 POST 请求的结果

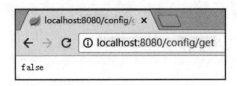

图 10-4　再次在浏览器中输入 localhost:8080/config/get 的结果

10.3　Nacos Discovery 的应用

视频讲解

10.3.1　服务提供者的实现

用 IDEA 创建完项目 nacosprovider 之后,确保在文件 pom.xml 的< dependencies >和</dependencies >之间添加了 Web、Nacos Discovery 依赖,代码如例 10-5 所示。

【例 10-5】　添加 Web、Nacos Discovery 依赖的代码示例。

```xml
<dependency>
        <groupId>org.springframework.boot</groupId>
        <artifactId>spring-boot-starter-web</artifactId>
</dependency>
<dependency>
        <groupId>org.springframework.cloud</groupId>
        <artifactId>spring-cloud-starter-alibaba-nacos-discovery</artifactId>
        <version>0.2.0.RELEASE</version>
</dependency>
```

在包 com.bookcode 中创建 controller 子包,并在包 com.bookcode.controller 中创建类 HelloController,代码如例 10-6 所示。

【例 10-6】　创建类 HelloController 的代码示例。

```java
package com.bookcode.controller;
import org.springframework.web.bind.annotation.PathVariable;
import org.springframework.web.bind.annotation.RequestMapping;
import org.springframework.web.bind.annotation.RequestMethod;
import org.springframework.web.bind.annotation.RestController;
@RestController
public class HelloController {
    @RequestMapping(value = "/hello/{string}", method = RequestMethod.GET)
    public String hello(@PathVariable String string) {
```

```
        return "Hello Nacos Discovery " + string;
    }
}
```

修改配置文件 application.properties，修改后的代码如例 10-7 所示。

【例 10-7】 修改后的配置文件 application.properties 的代码示例。

```
server.port = 18080
spring.application.name = service-provider
spring.cloud.nacos.discovery.server-addr = 127.0.0.1:8848
```

修改入口类，修改后的代码如例 10-8 所示。

【例 10-8】 修改后的入口类代码示例。

```
package com.bookcode;
import org.springframework.boot.SpringApplication;
import org.springframework.boot.autoconfigure.SpringBootApplication;
import org.springframework.cloud.client.discovery.EnableDiscoveryClient;
@SpringBootApplication
@EnableDiscoveryClient
public class DemoApplication {
    public static void main(String[] args) {
        SpringApplication.run(DemoApplication.class, args);
    }
}
```

10.3.2 服务消费者的实现

用 IDEA 创建完项目 nacosconsumer 之后，确保在文件 pom.xml 的 < dependencies > 和 </dependencies > 之间添加了 Web、Nacos Discovery 和 Ribbon 依赖，代码如例 10-9 所示。

【例 10-9】 添加 Web、Nacos Discovery 和 Ribbon 依赖的代码示例。

```xml
<dependency>
    <groupId>org.springframework.boot</groupId>
    <artifactId>spring-boot-starter-web</artifactId>
</dependency>
<dependency>
    <groupId>org.springframework.cloud</groupId>
    <artifactId>spring-cloud-starter-alibaba-nacos-discovery</artifactId>
    <version>0.2.0.RELEASE</version>
```

```
</dependency>
<dependency>
        <groupId>org.springframework.cloud</groupId>
        <artifactId>spring-cloud-starter-netflix-ribbon</artifactId>
</dependency>
```

在包 com.bookcode 中创建 controller 子包,并在包 com.bookcode.controller 中创建类 EchoController,代码如例 10-10 所示。

【例 10-10】 创建类 EchoController 的代码示例。

```
package com.bookcode.controller;
import org.springframework.beans.factory.annotation.Autowired;
import org.springframework.web.bind.annotation.PathVariable;
import org.springframework.web.bind.annotation.RequestMapping;
import org.springframework.web.bind.annotation.RequestMethod;
import org.springframework.web.bind.annotation.RestController;
import org.springframework.web.client.RestTemplate;
@RestController
public class EchoController {
    private final RestTemplate restTemplate;
    @Autowired
    public EchoController(RestTemplate restTemplate) {this.restTemplate = restTemplate;}
    @RequestMapping(value = "/echo/{str}", method = RequestMethod.GET)
    public String echo(@PathVariable String str) {
        return restTemplate.getForObject("http://service-provider/hello/" + str, String.class);
    }
}
```

修改配置文件 application.properties,修改后的代码如例 10-11 所示。

【例 10-11】 修改后的配置文件 application.properties 的代码示例。

```
spring.application.name = service-consumer
spring.cloud.nacos.discovery.server-addr = 127.0.0.1:8848
```

修改入口类,修改后的代码如例 10-12 所示。

【例 10-12】 修改后的入口类代码示例。

```
package com.bookcode;
import org.springframework.boot.SpringApplication;
import org.springframework.boot.autoconfigure.SpringBootApplication;
import org.springframework.cloud.client.discovery.EnableDiscoveryClient;
import org.springframework.cloud.client.loadbalancer.LoadBalanced;
```

```
import org.springframework.context.annotation.Bean;
import org.springframework.web.client.RestTemplate;
@SpringBootApplication
@EnableDiscoveryClient
public class DemoApplication {
    @LoadBalanced
    @Bean
    public RestTemplate restTemplate() {
        return new RestTemplate();
    }
    public static void main(String[] args) {
        SpringApplication.run(DemoApplication.class, args);
    }
}
```

10.3.3 运行程序

执行如例 10-4 所示命令,或者双击 startup.cmd 文件,启动 Nacos Server。

运行服务提供者的应用程序(nacosprovider),它的端口号为 18080。在浏览器中输入 localhost:18080/hello/zs,结果如图 10-5 所示。运行服务消费者的应用程序(nacosconsumer),它的端口为 8080。在浏览器中输入 localhost:8080/echo/ls,结果如图 10-6 所示。

图 10-5　在浏览器中输入 localhost: 18080/hello/zs 的结果

图 10-6　在浏览器中输入 localhost: 8080/echo/ls 的结果

10.4　Sentinel 的应用

视频讲解

随着微服务的流行,服务和服务之间的稳定性变得越来越重要。Sentinel 具有丰富的应用场景、完备的实时监控、广泛的开源生态、完善的服务提供者接口(Service Provider Interfaces,SPI)扩展点等特征。Sentinel 承接了阿里巴巴近 10 年的"双 11"大流量场景下的应用,例如秒杀(即突发流量控制在系统容量可以承受的范围)、消息削峰填谷、实时熔断下游不可用等应用。Sentinel 提供开箱即用与其他开源框架整合的模块,例如与 Spring Cloud、Dubbo 的整合。

10.4.1 创建项目并添加依赖

用 IDEA 创建完项目 sentinelexample 之后,确保在文件 pom.xml 的<dependencies>和</dependencies>之间添加了 Sentinel 和 Web 依赖,代码如例 10-13 所示。

【例 10-13】 添加 Sentinel 和 Web 依赖的代码示例。

```
<dependency>
        <groupId>org.springframework.cloud</groupId>
        <artifactId>spring-cloud-starter-alibaba-sentinel</artifactId>
        <version>0.2.0.RELEASE</version>
</dependency>
<dependency>
        <groupId>org.springframework.boot</groupId>
        <artifactId>spring-boot-starter-web</artifactId>
</dependency>
```

10.4.2 创建类 HelloController

在包 com.bookcode 中创建 controller 子包,并在包 com.bookcode.controller 中创建类 HelloController,代码如例 10-14 所示。

【例 10-14】 创建类 HelloController 的代码示例。

```
package com.bookcode.controller;
import com.alibaba.csp.sentinel.annotation.SentinelResource;
import org.springframework.web.bind.annotation.GetMapping;
import org.springframework.web.bind.annotation.PathVariable;
import org.springframework.web.bind.annotation.RestController;
@RestController
public class HelloController {
    @SentinelResource("resource")
    @GetMapping("/hello")
    public String hello() {
        return "您好!我是 Sentinel.";
    }
    @GetMapping(value = "/echo/{string}")
    public String echo(@PathVariable String string) {
```

```
            return "欢迎您," + string + "!";
    }
}
```

10.4.3　修改配置文件 application.properties

修改配置文件 application.properties,修改后的代码如例 10-15 所示。

【例 10-15】 修改后的配置文件 application.properties 的代码示例。

```
spring.application.name = sentinel-example
server.port = 18083
spring.cloud.sentinel.transport.dashboard = localhost:8080
```

10.4.4　运行程序

下载 Sentinel 控制台,执行如例 10-16 所示命令,启动 Sentinel 控制台。

【例 10-16】 启动 Sentinel 控制台的命令示例。

```
java -jar sentinel-dashboard.jar
```

运行程序 sentinelexample,在浏览器中输入 localhost:8080,自动跳转到 localhost:8080/#/dashboard/home,显示应用程序,结果如图 10-7 所示。在浏览器中输入 localhost:18083/echo/zs,结果如图 10-8 所示。在浏览器中输入 localhost:18083/hello,结果如图 10-9 所示。在浏览器中输入 localhost:8080/#/dashboard/identity/sentinel-example,结果如图 10-10 所示。

图 10-7　在浏览器中输入 localhost:8080 的结果

图 10-8　在浏览器中输入 localhost：18083/echo/zs 的结果

图 10-9　在浏览器中输入 localhost：18083/hello 的结果

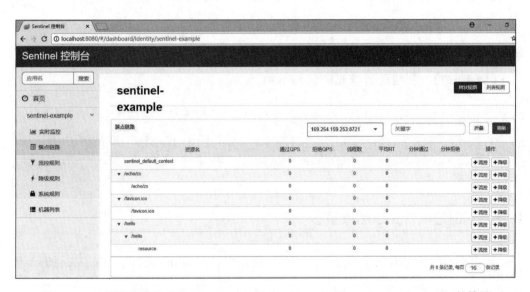

图 10-10　在浏览器中输入 localhost：8080/♯/dashboard/identity/sentinel-example 的结果

10.5　ACM 的应用

视频讲解

　　应用配置管理（Application Configuration Management，ACM）前身为淘宝内部配置中心 Diamond，是一款在分布式架构环境中对应用配置进行集中管理和推送的应用配置中心产品。利用 ACM，在微服务、DevOps（Development 和 Operations 的组合词）、大数据等场景下极大地减轻配置管理的工作量，并增强配置管理的服务能力。在应用生命周期管理中，开发人员通常会将应用中需要变更的一些配置项或者元数据从代码中分离出来，放在单独的配置文件中管理。

　　发布应用后，运维人员或最终用户可以通过调整配置来适配环境，或调整应用程序的运行行为。ACM 是面向分布式系统的配置中心。凭借配置变更、配置推送、历史版本管理、灰度发布、配置变更审计等配置管理工具,ACM 能集中管理所有应用环境中的配置,降低管理配置的成本,并降低因错误的配置变更带来可用性下降甚至发生故障的风险。

10.5.1 辅助工作

首先,在阿里云注册用户,获得个人的 accessKey 和 secretKey。接着,开通应用配置管理(ACM)。然后,前往 ACM 控制台,创建命名空间(namespace);在相应的命名空间下新建相关配置信息,如例10-17所示。

【例10-17】 新建配置信息的代码示例。

```
user.id = 001
user.name = 张三
user.age = 28
```

10.5.2 创建项目并添加依赖

用 IDEA 创建完项目 acmexample 之后,确保在文件 pom.xml 的< dependencies >和</dependencies >之间添加了 ACM、Actuator 和 Web 依赖,代码如例10-18所示。

【例10-18】 添加 ACM、Actuator 和 Web 依赖的代码示例。

```xml
<dependency>
    <groupId>com.alibaba.cloud</groupId>
    <artifactId>spring-cloud-starter-acm</artifactId>
    <version>1.0.7</version>
</dependency>
<dependency>
    <groupId>org.springframework.boot</groupId>
    <artifactId>spring-boot-starter</artifactId>
</dependency>
<dependency>
    <groupId>org.springframework.boot</groupId>
    <artifactId>spring-boot-starter-actuator</artifactId>
</dependency>
<dependency>
    <groupId>org.springframework.boot</groupId>
    <artifactId>spring-boot-starter-web</artifactId>
</dependency>
```

10.5.3 创建类 SampleController

在包 com.bookcode 中创建 controller 子包,并在包 com.bookcode.controller 中创建类 SampleController,代码如例 10-19 所示。

【例 10-19】 创建类 SampleController 的代码示例。

```
package com.bookcode.controller;
import org.springframework.beans.factory.annotation.Value;
import org.springframework.cloud.context.config.annotation.RefreshScope;
import org.springframework.web.bind.annotation.RequestMapping;
import org.springframework.web.bind.annotation.RestController;
@RestController
@RequestMapping("/scalibaba")
@RefreshScope
class SampleController {
    @Value("${user.name}")
    String userName;
    @Value("${user.id}")
    String userId;
    @RequestMapping("/acm")
    public String simple() {
        return "欢迎您 " + userId + userName + "!我是阿里 Spring Cloud 应用配置管理器(ACM)" + ",当前控制器是" + this + ".";
    }
}
```

10.5.4 修改配置文件 application.properties

修改配置文件 application.properties,修改后的代码如例 10-20 所示。

【例 10-20】 修改后的配置文件 application.properties 的代码示例。

```
spring.application.group = com.alibaba.cloud.acm
spring.application.name = sample-app
alibaba.acm.endpoint = acm.aliyun.com
alibaba.acm.namespace = baf7ea46-6ef0-4a20-9e0e-d23e6223afbc
alibaba.acm.accessKey = LTAIbbIdpGnFczr6
alibaba.acm.secretKey = wzsoySTa0fxHyXi0tCg16bF7OQ3YYz
management.endpoints.web.exposure.include = *
```

10.5.5 运行程序

运行程序,在浏览器中输入 localhost:8080/scalibaba/acm,结果如图 10-11 所示。

图 10-11　在浏览器中输入 localhost:8080/scalibaba/acm 的结果

习题 10

一、问答题

1. 请简述 Spring Cloud Alibaba 的主要功能。
2. 请简述 Spring Cloud Alibaba 的主要组件。

二、实验题

1. 请实现 Nacos Config 的简单应用。
2. 请实现 Nacos Discovery 的简单应用。
3. 请实现 Sentinel 的应用。
4. 请实现 ACM 的应用。

第 11 章

Dubbo 的应用

本章介绍 Dubbo、Dubbo Spring Boot 和 Spring Cloud Dubbo 的应用。Dubbo 是阿里巴巴公司开源的一款高性能、轻量级的 Java RPC(Remote Procedure Call,远程过程调用)框架。Dubbo 作为成熟的 RPC 框架,其易用性、扩展性和健壮性已得到业界的认可。Dubbo 将会作为 Spring Cloud Alibaba 的 RPC 组件,并与 Spring Cloud 原生的 Feign 以及 RestTemplate 进行无缝整合,实现"零"成本迁移。

11.1 Dubbo 简介

11.1.1 Dubbo 主要功能

Dubbo 提供了三大核心能力:面向接口的远程方法调用,智能容错和负载均衡以及服务自动注册和发现。除此之外,还包括统计服务的调用次数和调用时间的监控中心。

远程方法调用功能提供对多种基于长连接的 NIO(Non-Blocking IO,非阻塞 IO)框架抽象封装,包括多种线程模型、序列化以及"请求-响应"模式的信息交换方式。

集群容错功能提供基于接口方法的透明远程过程调用,包括多协议支持以及软负载均衡、失败容错、地址路由、动态配置等集群支持。

服务自动发现功能基于注册中心目录服务,使服务消费方能动态地查找服务提供方,使地址透明,使服务提供方可以平滑增加或减少机器。

11.1.2　Dubbo Spring Boot 简介

Dubbo Spring Boot 工程致力于简化 Dubbo 框架在 Spring Boot 应用中的开发。同时也整合了 Spring Boot 自动装配、注解驱动、安全、健康检查、外部化配置等特性。

Dubbo Spring Boot 采用多 Maven 模块工程,通常包括 dubbo-spring-boot-parent、dubbo-spring-boot-autoconfigure、dubbo-spring-boot-actuator、dubbo-spring-boot-starter 等模块。dubbo-spring-boot-parent 模块主要管理 Dubbo Spring Boot 工程的 Maven 依赖。dubbo-spring-boot-autoconfigure 模块提供 Spring Boot 的注解@EnableAutoConfiguration 的实现(DubboAutoConfiguration),它简化了 Dubbo 核心组件的装配。dubbo-spring-boot-actuator 提供 Production-Ready 特性(健康检查、控制断点、外部化配置)。dubbo-spring-boot-starter 模块为标准的 Spring Boot 开箱即用模块,将它引入到工程后,dubbo-spring-boot-autoconfigure 模块会一同被间接依赖。

11.2　Dubbo 的简单应用

视频讲解

11.2.1　服务提供者的实现

用 IDEA 创建完项目 dubboprovider 之后,确保在文件 pom.xml 的< dependencies >和</dependencies >之间添加了 Dubbo、Zookeeper Client 和 Web 依赖,代码如例 11-1 所示。

【例 11-1】　添加 Dubbo、Zookeeper Client 和 Web 依赖的代码示例。

```
<dependency>
    <groupId>com.alibaba.boot</groupId>
    <artifactId>dubbo-spring-boot-starter</artifactId>
```

```xml
            <version>0.2.0</version>
</dependency>
<!-- 引入 ZookeeperClient -->
<dependency>
            <groupId>com.github.sgroschupf</groupId>
            <artifactId>zkclient</artifactId>
            <version>0.1</version>
</dependency>
<dependency>
            <groupId>org.springframework.boot</groupId>
            <artifactId>spring-boot-starter-web</artifactId>
</dependency>
```

在包 com.bookcode 中创建 service 子包，并在包 com.bookcode.service 中创建接口 HelloService，代码如例 11-2 所示。

【例 11-2】 创建接口 HelloService 的代码示例。

```java
package com.bookcode.service;
public interface HelloService {
    String sayHello(String name);
    String hello();
}
```

在包 com.bookcode.service 中创建 impl 子包，并在包 com.bookcode.service.impl 中创建类 HelloServiceImpl，代码如例 11-3 所示。

【例 11-3】 创建类 HelloServiceImpl 的代码示例。

```java
package com.bookcode.service.impl;
import com.alibaba.dubbo.config.annotation.Service;
import com.bookcode.service.HelloService;
 @Service//是 Dubbo 注解而不是 Spring 注解
public class HelloServiceImpl implements HelloService {
    @Override
     public String sayHello(String name) {
        return "Hello, " + name + " (from Spring Boot)";
    }
    @Override
    public String hello() {
        return "您好," + "我是 Dubbo。";
    }
}
```

在包 com.bookcode 中创建 controller 子包，并在包 com.bookcode.controller 中

创建类 EchoController,代码如例 11-4 所示。

【例 11-4】 创建类 EchoController 的代码示例。

```
package com.bookcode.controller;
import org.springframework.web.bind.annotation.GetMapping;
import org.springframework.web.bind.annotation.PathVariable;
import org.springframework.web.bind.annotation.RestController;
@RestController
public class EchoController {
    @GetMapping(value = "/echo/{string}")
    public String echo(@PathVariable String string) {
        return "欢迎您," + string + "!";
    }
}
```

修改目录 src/main/resources 下配置文件 application.properties,修改后的代码如例 11-5 所示。

【例 11-5】 修改后的配置文件 application.properties 的代码示例。

```
server.port = 9000
```

在目录 src/main/resources 下,创建并修改配置文件 application-spring.xml,修改后的代码如例 11-6 所示。

【例 11-6】 修改后的配置文件 application-spring.xml 的代码示例。

```xml
<?xml version = "1.0" encoding = "UTF-8"?>
<beans xmlns = "http://www.springframework.org/schema/beans"
       xmlns:xsi = "http://www.w3.org/2001/XMLSchema-instance"
       xmlns:dubbo = "http://code.alibabatech.com/schema/dubbo"
       xsi:schemaLocation = "http://www.springframework.org/schema/beans
          http://www.springframework.org/schema/beans/spring-beans.xsd
          http://code.alibabatech.com/schema/dubbo
          http://code.alibabatech.com/schema/dubbo/dubbo.xsd">
    <dubbo:application name = "hello-world-app1"/>
    <dubbo:registry protocol = "zookeeper" address = "127.0.0.1:2181" />
    <dubbo:protocol name = "dubbo" port = "20881" />
    <!-- <Dubbo:annotation>,是 dubbo 的扫描标签,它除了会扫描带有
        @Component、@Service、@Controller 注解的类,把它们注册成 SpringBean 之外,
        它还会扫描带有@Service(dubbo)的接口实现类发布服务
        (必须有实现接口,不然或抛出 BeanCreationException 异常)。
        同时在要发布服务的接口实现类上加上@Service(dubbo)。
        启动服务器,服务就发布成功了。-->
    <dubbo:annotation package = "com.bookcode.service.impl"></dubbo:annotation>
</beans>
```

修改入口类,修改后的代码如例 11-7 所示。

【例 11-7】 修改后的入口类的代码示例。

```
package com.bookcode;
import org.springframework.boot.SpringApplication;
import org.springframework.boot.autoconfigure.SpringBootApplication;
import org.springframework.context.annotation.ImportResource;
@SpringBootApplication()
@ImportResource("classpath:application-spring.xml")
public class DemoApplication {
    public static void main(String[] args) {
        SpringApplication.run(DemoApplication.class, args);
    }
}
```

11.2.2 服务消费者的实现

用 IDEA 创建完项目 dubboconsumer 之后,确保在文件 pom.xml 的<dependencies>和</dependencies>之间添加了 Dubbo、Zookeeper Client 和 Web 依赖,代码如例 11-1 所示。

在包 com.bookcode 中创建 service 子包,并在包 com.bookcode.service 中创建接口 HelloService,代码如例 11-2 所示。

在包 com.bookcode 中创建 controller 子包,并在包 com.bookcode.controller 中创建类 HelloController,代码如例 11-8 所示。

【例 11-8】 创建类 HelloController 的代码示例。

```
package com.bookcode.controller;
import com.alibaba.dubbo.config.annotation.Reference;
import com.bookcode.service.HelloService;
import org.springframework.web.bind.annotation.GetMapping;
import org.springframework.web.bind.annotation.PathVariable;
import org.springframework.web.bind.annotation.RestController;
@RestController
public class HelloController {
    @Reference
    private HelloService helloService;
    @GetMapping(value = "/hi")
    public String hi(){
        return helloService.hello();
    }
    @GetMapping(value = "/greet/{string}")
```

```
    public String greet(@PathVariable String string) {
        return helloService.sayHello(string);
    }
}
```

修改目录 src/main/resources 下配置文件 application.properties,修改后的代码如例 11-9 所示。

【例 11-9】 修改后的配置文件 application.properties 的代码示例。

```
# 避免和 server 工程端口冲突
server.port = 9080
dubbo.application.name = dubbo-consumer-demo
dubbo.registry.address = zookeeper://127.0.0.1:2181
dubbo.scan = com.bookcode
```

11.2.3 运行程序

执行如例 11-10 所示命令,启动 Zookeeper 服务。

【例 11-10】 启动 Zookeeper 服务的命令示例。

```
zkServer
```

依次运行服务提供者的应用程序 dubboprovider,它的端口为 9000。运行服务消费者的应用程序 dubboconsumer,它的端口为 9080。在浏览器中输入 localhost:9000/echo/Dubbo,结果如图 11-1 所示。在浏览器中输入 localhost:9080/hi,结果如图 11-2 所示。在浏览器中输入 localhost:9080/greet/Dubbo-Alibaba,结果如图 11-3 所示。

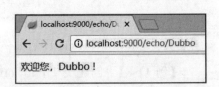

图 11-1 在浏览器中输入 localhost:9000/echo/Dubbo 的结果

图 11-2 在浏览器中输入 localhost:9080/hi 的结果

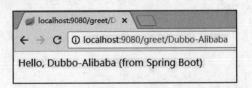

图 11-3 在浏览器中输入 localhost:9080/greet/Dubbo-Alibaba 的结果

11.3 Dubbo Spring Boot 的应用

视频讲解

11.3.1 服务提供者的实现

用 IDEA 创建完项目 dubboprovidernew 之后，确保在文件 pom.xml 的 <dependencies>和</dependencies>之间添加了 Dubbo Spring Boot 和 Web 依赖，代码如例 11-11 所示。

注意，对比例 11-1 第 2 行代码和例 11-11 第 2 行代码的不同。

【例 11-11】 添加 Dubbo Spring Boot 和 Web 依赖的代码示例。

```
<dependency>
    <groupId>com.alibaba.spring.boot</groupId>
    <artifactId>dubbo-spring-boot-starter</artifactId>
    <version>2.0.0</version>
</dependency>
<dependency>
    <groupId>org.springframework.boot</groupId>
    <artifactId>spring-boot-starter-web</artifactId>
</dependency>
```

在包 com.bookcode 中创建 service 子包，并在包 com.bookcode.service 中创建接口 AddService，代码如例 11-12 所示。

【例 11-12】 创建接口 AddService 的代码示例。

```
package com.bookcode.service;
public interface AddService {
    int add(int x, int y);
}
```

在包 com.bookcode.service 中创建 impl 子包，并在包 com.bookcode.service.impl 中创建类 AddServiceImpl，代码如例 11-13 所示。

【例 11-13】 创建类 AddServiceImpl 的代码示例。

```
package com.bookcode.service.impl;
import com.alibaba.dubbo.config.annotation.Service;
import com.bookcode.service.AddService;
import org.springframework.stereotype.Component;
```

```
@Service(interfaceClass = AddService.class)//是Dubbo的注解而不是Spring注解
@Component
public class AddServiceImpl implements AddService {
    @Override
    public int add(int x, int y) {
        return x + y;
    }
}
```

在包com.bookcode中创建controller子包,并在包com.bookcode.controller中创建类SubController,代码如例11-14所示。

【例11-14】 创建类SubController的代码示例。

```
package com.bookcode.controller;
import org.springframework.web.bind.annotation.GetMapping;
import org.springframework.web.bind.annotation.PathVariable;
import org.springframework.web.bind.annotation.RestController;
@RestController
public class SubController {
    @GetMapping(value = "/sub/{x}/{y}")
    public String sub(@PathVariable int x,@PathVariable int y) {
        return String.valueOf(x) + " - " + String.valueOf(y) + " = " + String.valueOf(x-y);
    }
}
```

修改目录src/main/resources下配置文件application.properties,修改后的代码如例11-15所示。

【例11-15】 修改后的配置文件application.properties的代码示例。

```
spring.application.name = dubbo-spring-boot-starter-example-provider
spring.dubbo.server = true
spring.dubbo.registry = N/A
server.port = 8790
```

修改入口类,修改后的代码如例11-16所示。

【例11-16】 修改后的入口类的代码示例。

```
package com.bookcode;
import com.alibaba.dubbo.spring.boot.annotation.EnableDubboConfiguration;
import org.springframework.boot.SpringApplication;
import org.springframework.boot.autoconfigure.SpringBootApplication;
@SpringBootApplication
@EnableDubboConfiguration
```

```
public class DemoApplication {
    public static void main(String[] args) {
        SpringApplication.run(DemoApplication.class, args);
    }
}
```

11.3.2 服务消费者的实现

用 IDEA 创建完项目 dubboconsumernew 之后,确保在文件 pom.xml 的 <dependencies>和</dependencies>之间添加了 Dubbo Spring Boot 和 Web 依赖,代码如例 11-11 所示。

在包 com.bookcode 中创建 service 子包,并在包 com.bookcode.service 中创建接口 AddService,代码如例 11-12 所示。

在包 com.bookcode 中创建 controller 子包,并在包 com.bookcode.controller 中创建类 AddController,代码如例 11-17 所示。

【例 11-17】 创建类 AddController 的代码示例。

```
package com.bookcode.controller;
import com.alibaba.dubbo.config.annotation.Reference;
import com.bookcode.service.AddService;
import org.springframework.web.bind.annotation.GetMapping;
import org.springframework.web.bind.annotation.PathVariable;
import org.springframework.web.bind.annotation.RestController;
@RestController
public class AddController {
    @Reference(url = "dubbo://127.0.0.1:20880")
    private AddService addService;
    @GetMapping(value = "/add/{x}/{y}")
    public String add(@PathVariable int x, @PathVariable int y) {
        return String.valueOf(x) + " + " + String.valueOf(y) + " = " + String.valueOf(addService.add(x,y));
    }
}
```

修改目录 src/main/resources 下配置文件 application.properties,修改后的代码如例 11-18 所示。

【例 11-18】 修改后的配置文件 application.properties 的代码示例。

```
spring.application.name = dubbo-spring-boot-starter-example-consumer
server.port = 8794
```

修改入口类，修改后的代码如例 11-16 所示。

11.3.3 运行程序

依次运行服务提供者的应用程序 dubboprovidernew，它的端口为 8790。运行服务消费者的应用程序 dubboconsumernew，它的端口为 8794。在浏览器中输入 localhost:8790/sub/7/2，结果如图 11-4 所示。在浏览器中输入 localhost:8794/add/5/2，结果如图 11-5 所示。

图 11-4 在浏览器中输入 localhost：
8790/sub/7/2 的结果

图 11-5 在浏览器中输入 localhost：
8794/add/5/2 的结果

11.4 Spring Cloud Dubbo 的应用

视频讲解

11.4.1 服务提供者的实现

用 IDEA 创建完项目 clouddubboprovider 之后，确保在文件 pom.xml 的 <dependencies> 和 </dependencies> 之间添加了 Security、Eureka、Client、Dubbo、Openfeign、Web 依赖，代码如例 11-19 所示。

【例 11-19】 添加 Security、Eureka、Client、Dubbo、Openfeign、Web 依赖的代码示例。

```xml
<dependency>
    <groupId>org.springframework.cloud</groupId>
    <artifactId>spring-cloud-starter-security</artifactId>
</dependency>
<dependency>
    <groupId>org.springframework.cloud</groupId>
    <artifactId>spring-cloud-starter-netflix-eureka-client</artifactId>
</dependency>
<dependency>
    <groupId>cn.springcloud.dubbo</groupId>
```

```xml
        <artifactId>spring-cloud-dubbo-starter</artifactId>
</dependency>
<dependency>
        <groupId>com.alibaba.spring.boot</groupId>
        <artifactId>dubbo-spring-boot-starter</artifactId>
        <version>2.0.0</version>
</dependency>
<dependency>
        <groupId>org.springframework.cloud</groupId>
        <artifactId>spring-cloud-openfeign-core</artifactId>
</dependency>
<dependency>
        <groupId>org.springframework.boot</groupId>
        <artifactId>spring-boot-starter-web</artifactId>
</dependency>
```

在包 com.bookcode 中创建 service 子包,并在包 com.bookcode.service 中创建接口 HelloService,代码如例 11-20 所示。

【例 11-20】 创建接口 HelloService 的代码示例。

```java
package com.bookcode.service;
import org.springframework.cloud.openfeign.FeignClient;
import org.springframework.web.bind.annotation.GetMapping;
@FeignClient("provider")
public interface HelloService {
    @GetMapping("/hello")
    String hello();
}
```

在包 com.bookcode.service 中创建类 HelloServiceImpl,代码如例 11-21 所示。

【例 11-21】 创建类 HelloServiceImpl 的代码示例。

```java
package com.bookcode.service;
import com.alibaba.dubbo.config.annotation.Service;
import org.springframework.web.bind.annotation.RestController;
@RestController
@Service(interfaceClass = HelloService.class)
public class HelloServiceImpl implements HelloService {
    @Override
    public String hello() {
        return "hello at " + System.currentTimeMillis();
    }
}
```

修改目录 src/main/resources 下配置文件 application.properties，修改后的代码如例 11-22 所示。

【例 11-22】 修改后的配置文件 application.properties 的代码示例。

```
server.port = 8790
spring.application.name = provider
#开启基于 HTTP basic 的认证
spring.security.basic.enable = true
spring.security.user.name = zs
spring.security.user.password = zs
spring.dubbo.server = true
spring.dubbo.registry = N/A
eureka.client.serviceUrl.defaultZone = http://localhost:8761/eureka/
info.app.name = @project.artifactId@
info.encoding = @project.build.sourceEncoding@
info.java.source = @java.version@
```

修改入口类，修改后的入口类代码如例 11-23 所示。

【例 11-23】 修改后的入口类的代码示例。

```
package com.bookcode;
import com.alibaba.dubbo.spring.boot.annotation.EnableDubboConfiguration;
import org.springframework.boot.SpringApplication;
import org.springframework.boot.autoconfigure.SpringBootApplication;
@SpringBootApplication
@EnableDubboConfiguration
public class DemoApplication {
    public static void main(String[] args) {
        SpringApplication.run(DemoApplication.class, args);
    }
}
```

11.4.2 服务消费者的实现

用 IDEA 创建完项目 clouddubboconsumer 之后，确保在文件 pom.xml 的 <dependencies> 和 </dependencies> 之间添加了 Security、Eureka、Client、Dubbo、Openfeign、Web 等依赖，代码如例 11-20 所示。

在包 com.bookcode 中创建 service 子包，并在包 com.bookcode.service 中创建接口 HelloService，代码如例 11-20 所示。在包 com.bookcode.service 中创建类 TestServiceController，代码如例 11-24 所示。

【例 11-24】 创建类 TestServiceController 的代码示例。

```
package com.bookcode.service;
import com.alibaba.dubbo.config.annotation.Reference;
import org.springframework.web.bind.annotation.GetMapping;
import org.springframework.web.bind.annotation.RestController;
@RestController
public class TestServiceController {
    @Reference(url = "dubbo://127.0.0.1:20880")
    private HelloService helloService;
    @GetMapping("/test")
    public String test() {
        return "Test " + helloService.hello();
    }
}
```

修改目录 src/main/resources 下配置文件 application.properties,修改后的代码如例 11-25 所示。

【例 11-25】 修改后的配置文件 application.properties 的代码示例。

```
server.port = 9000
spring.application.name = consumer
eureka.client.serviceUrl.defaultZone = http://localhost:8761/eureka/
info.app.name = @project.artifactId@
info.encoding = @project.build.sourceEncoding@
info.java.source = @java.version@
```

修改入口类,修改后的代码如例 11-23 所示。

11.4.3 运行程序

依次运行 3.2 节 eureka-server 程序,它的端口为 8761;运行服务提供者的应用程序 clouddubboprovider,它的端口为 8790;运行服务消费者的应用程序 clouddubboconsumer,它的端口为 9000。在浏览器中输入 localhost:8761,结果如图 11-6 所示。单击应用 PROVIDER,结果跳转到 localhost:8790/login,结果如图 11-7 所示。输入正确的 User 和 Password 后,结果跳转到 localhost:8790/actuator/info,结果如图 11-8 所示。在浏览器中输入 localhost:8790/hello,结果如图 11-9 所示。在浏览器中输入 localhost:9000/test,结果如图 11-10 所示。

Application	AMIs	Availability Zones	Status
CONSUMER	n/a (1)	(1)	UP (1) - LAPTOP-H5IO2QIN:consumer:9000
PROVIDER	n/a (1)	(1)	UP (1) - LAPTOP-H5IO2QIN:provider:8790

图 11-6 在浏览器中输入 localhost:8761 的结果

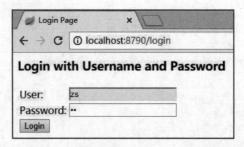

图 11-7　跳转到 localhost:8790/login 的结果

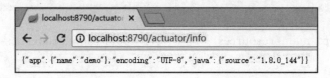

图 11-8　跳转到 localhost:8790/actuator/info 的结果

 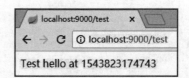

图 11-9　在浏览器中输入 localhost: 8790/hello 的结果

图 11-10　在浏览器中输入 localhost: 9000/test 的结果

习题 11

一、问答题

请简述 Dubbo 的主要功能。

二、实验题

1．请实现 Dubbo 的简单应用。

2．请实现 Dubbo Spring Boot 的应用。

3．请实现 Spring Cloud Dubbo 的应用。

第 12 章

Spring Cloud 的综合应用案例

本章结合一个简单案例介绍 Spring Cloud 的应用开发。

视频讲解

12.1 实现配置中心 case-config-server

12.1.1 创建项目并添加依赖

用 IDEA 创建完项目 taskexample 之后,确保在文件 pom.xml 的<dependencies>和</dependencies>之间添加了 Config Server 依赖,代码如例 12-1 所示。

【例 12-1】 添加 Config Server 依赖的代码示例。

```
<dependency>
    <groupId>org.springframework.cloud</groupId>
    <artifactId>spring-cloud-config-server</artifactId>
</dependency>
```

12.1.2 创建配置文件 application.yml

在目录 src/main/resources 下,创建配置文件 application.yml,代码如例 12-2 所示。

【例12-2】 创建的配置文件 application.yml 代码示例。

```yaml
eureka:
  client:
    register-with-eureka: false
    fetch-registry: false
server:
  port: 8040
spring:
  application:
    name: microservice-config-server
  cloud:
    config:
      server:
        git:
          uri: https://github.com/woodsheng/config-repo.git
          search-paths: config-repo
          username:
          password:
```

12.1.3 修改入口类

修改入口类，修改后的代码如例12-3所示。

【例12-3】 修改后的入口类代码示例。

```java
package com.bookcode;
import org.springframework.boot.SpringApplication;
import org.springframework.boot.autoconfigure.SpringBootApplication;
import org.springframework.cloud.config.server.EnableConfigServer;
@EnableConfigServer
@SpringBootApplication
public class DemoApplication {
    public static void main(String[] args) {
        SpringApplication.run(DemoApplication.class, args);
    }
}
```

12.1.4 运行程序

依次运行3.2节中 eureka-server 程序和本节的 case-config-server 程序。在浏览器中输入 localhost:8040/microservice-config-client/dev，结果如图12-1所示。在浏

览器中输入 localhost:8040/microservice-config-client-dev.properties，结果如图 12-2 所示。

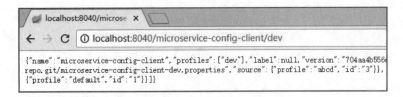

图 12-1　在浏览器中输入 localhost:8040/microservice-config-client/dev 的结果

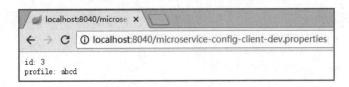

图 12-2　在浏览器中输入 localhost:8040/microservice-config-client-dev.properties 的结果

12.2　实现客户端服务 case-eureka-user-client

视频讲解

配置中心搭建完成后，接下来搭建一个客户端，该客户端需要用到配置中心的配置信息。

12.2.1　创建项目并添加依赖

用 IDEA 创建完项目 case-eureka-user-client 之后，确保在文件 pom.xml 的 <dependencies>和</dependencies>之间添加了 Eureka Client 等依赖，代码如例 12-4 所示。

【例 12-4】　添加 Eureka Client 等依赖的代码示例。

```
<dependency>
    <groupId>org.springframework.cloud</groupId>
    <artifactId>spring-cloud-starter-netflix-eureka-client</artifactId>
</dependency>
<dependency>
    <groupId>org.springframework.boot</groupId>
    <artifactId>spring-boot-starter-data-jpa</artifactId>
```

```xml
        </dependency>
        <dependency>
            <groupId>org.springframework.cloud</groupId>
            <artifactId>spring-cloud-starter-config</artifactId>
        </dependency>
        <dependency>
            <groupId>org.springframework.boot</groupId>
            <artifactId>spring-boot-starter-web</artifactId>
        </dependency>
        <dependency>
            <groupId>mysql</groupId>
            <artifactId>mysql-connector-java</artifactId>
        </dependency>
```

12.2.2 创建类 User

在包 com.bookcode 中创建 pojo 子包,并在包 com.bookcode.pojo 中创建类 User,代码如例 12-5 所示。

【例 12-5】 创建类 User 的代码示例。

```java
package com.bookcode.pojo;
import javax.persistence.*;
@Entity
@Table(name = "user")
public class User {
    @Id                              //@Id指明将属性id映射为数据库的主键
    @GeneratedValue(strategy = GenerationType.IDENTITY)    //采用自增策略增加id值
    private Long id;
    private String firstName;
    private String lastName;
    protected User() {}
    public Long getId() {
        return id;
    }
    public void setId(Long id) {
        this.id = id;
    }
    public String getFirstName() {
        return firstName;
    }
    public void setFirstName(String firstName) {
        this.firstName = firstName;
    }
    public String getLastName() {
```

```
        return lastName;
    }
    public void setLastName(String lastName) {
        this.lastName = lastName;
    }
    public User(String firstName, String lastName) {
        this.firstName = firstName;
        this.lastName = lastName;
    }
    @Override
    public String toString() {
        return String.format(
                "User[id=%d, firstName='%s', lastName='%s']",
                id, firstName, lastName);
    }
}
```

12.2.3 创建接口 UserDao

在包 com.bookcode 中创建 dao 子包,并在包 com.bookcode.dao 中创建接口 UserDao,代码如例 12-6 所示。

【例 12-6】 创建接口 UserDao 的代码示例。

```
package com.bookcode.dao;
import com.bookcode.pojo.User;
import org.springframework.data.jpa.repository.JpaRepository;
public interface UserDao extends JpaRepository<User, Long> {
}
```

12.2.4 创建类 UserController

在包 com.bookcode 中创建子包 controller,并在包 com.bookcode.controller 中创建类 UserController,代码如例 12-7 所示。

【例 12-7】 创建类 UserController 的代码示例。

```
package com.bookcode.controller;
import com.bookcode.dao.UserDao;
import com.bookcode.pojo.User;
import org.springframework.beans.factory.annotation.Autowired;
import org.springframework.beans.factory.annotation.Value;
```

```java
import org.springframework.web.bind.annotation.*;
@RestController
public class UserController {
    @Value("${id}")
    private Long id;
    @Autowired
    private UserDao userDao;
    @RequestMapping("/hi")
    public String hi() {
        User user = userDao.getOne(id);
        if(user == null)
        {
            user = new User("Error", "Error");
        }
        return user.toString();
    }
    @RequestMapping("getUserById")
    public User getUserById (@RequestParam Long id){
        User user = userDao.getOne(id);
        if(user == null)
        {
            user = new User("Error", "Error");
        }
        return user;
    }
}
```

12.2.5 修改和创建配置文件

修改目录 src/main/resources 下的配置文件 application.properties，修改后的代码如例 12-8 所示。

【例 12-8】 修改后的配置文件 application.properties 的代码示例。

```
spring.datasource.driverClassName = com.mysql.jdbc.Driver
spring.datasource.url = jdbc:mysql://localhost:3306/mytest?useUnicode=true&characterEncoding=utf-8
spring.datasource.username = root
spring.datasource.password = sa
spring.jackson.serialization.fail-on-empty-beans = false
```

在 src/main/resources 目录下，创建配置文件 application.yml，代码如例 12-9 所示。

【例 12-9】 创建的配置文件 application.yml 的代码示例。

```yaml
eureka:
  client:
    serviceUrl:
      defaultZone: http://localhost:8761/eureka/
server:
  port: 8091
spring:
  application:
    name: service-user
```

在目录 src/main/resources 下,创建配置文件 bootstrap.yml,代码如例 12-10 所示。

【例 12-10】 创建的配置文件 bootstrap.yml 代码示例。

```yaml
spring:
  cloud:
    config:
      name: microservice-config-client
      uri: http://localhost:8040
      profile: dev
      label: master
```

12.2.6 修改入口类

修改入口类,修改后的代码如例 12-11 所示。

【例 12-11】 修改后的入口类代码示例。

```java
package com.bookcode;
import com.bookcode.dao.UserDao;
import com.bookcode.pojo.User;
import org.slf4j.Logger;
import org.slf4j.LoggerFactory;
import org.springframework.boot.CommandLineRunner;
import org.springframework.boot.SpringApplication;
import org.springframework.boot.autoconfigure.SpringBootApplication;
import org.springframework.cloud.client.discovery.EnableDiscoveryClient;
import org.springframework.context.annotation.Bean;
@EnableDiscoveryClient
@SpringBootApplication
```

```
public class DemoApplication {
    private static final Logger log = LoggerFactory.getLogger(DemoApplication.class);
    @Bean
    public CommandLineRunner demo(UserDao repository) {
        return (args) -> {
            //存储 5 条用户记录到数据库
            repository.save(new User("Jack", "Bauer"));
            repository.save(new User("Chloe", "O'Brian"));
            repository.save(new User("Kim", "Bauer"));
            repository.save(new User("David", "Palmer"));
            repository.save(new User("Michelle", "Dessler"));
            //将所有 User 信息以日志形式输出到控制台
            log.info("Users found with findAll():");
            log.info("-------------------------------");
            for (Object user : repository.findAll()) {
                log.info(user.toString());            }
            //获取 id = 1 的 user 信息,并以日志形式在控制台上输出
            repository.findById(1L)
                    .ifPresent(User -> {
                        log.info("User found with findById(1L):");
                        log.info("-------------------------------");
                        log.info(User.toString());
                    });
        };
    }
    public static void main(String[] args) {
        SpringApplication.run(DemoApplication.class, args);
    }
}
```

12.2.7 运行程序

在运行 3.3 节 eureka-server 程序和 12.1 节 case-config-server 程序的基础上,运行本节 case-eureka-user-client 程序。在浏览器中输入 localhost:8091/hi,结果如图 12-3 所示。在浏览器中输入 localhost:8091/getUserById? id=2,结果如图 12-4 所示。

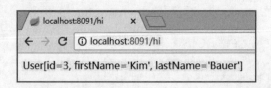

图 12-3 在浏览器中输入 localhost:8091/hi 的结果

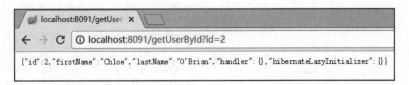

图 12-4　在浏览器中输入 localhost:8091/getUserById? id＝2 的结果

12.3　实现服务消费端 case-user-ribbon

视频讲解

12.3.1　创建项目并添加依赖

用 IDEA 创建完项目 case-user-ribbon 之后,确保在文件 pom.xml 的＜dependencies＞和＜/dependencies＞之间添加了 Ribbon 等依赖,代码如例 12-12 所示。

【例 12-12】　添加 Ribbon 等依赖的代码示例。

```
<dependency>
        <groupId>org.springframework.cloud</groupId>
        <artifactId>spring-cloud-starter-netflix-ribbon</artifactId>
</dependency>
<dependency>
        <groupId>org.springframework.cloud</groupId>
        <artifactId>spring-cloud-starter-hystrix</artifactId>
</dependency>
<dependency>
        <groupId>org.springframework.cloud</groupId>
        <artifactId>spring-cloud-starter-netflix-eureka-client</artifactId>
</dependency>
<dependency>
        <groupId>org.springframework.boot</groupId>
        <artifactId>spring-boot-starter-web</artifactId>
</dependency>
```

12.3.2　创建类 User

在包 com.bookcode 中创建 pojo 子包,并在包 com.bookcode.pojo 中创建类 User,代码如例 12-13 所示。

【例 12-13】　创建类 User 的代码示例。

```java
package com.bookcode.pojo;
public class User {
    public Long getId() {
        return id;
    }
    public void setId(Long id) {
        this.id = id;
    }
    public String getFirstName() {
        return firstName;
    }
    public void setFirstName(String firstName) {
        this.firstName = firstName;
    }
    public String getLastName() {
        return lastName;
    }
    public void setLastName(String lastName) {
        this.lastName = lastName;
    }
    private Long id;
    private String firstName;
    private String lastName;
    protected User() {}
    public User(String firstName, String lastName) {
        this.firstName = firstName;
        this.lastName = lastName;
    }
    @Override
    public String toString() {
        return String.format(
                "User[id=%d, firstName='%s', lastName='%s']",
                id, firstName, lastName);
    }
}
```

12.3.3 创建类 UserRibbonService

在包 com.bookcode 中创建 service 子包,并在包 com.bookcode.service 中创建类 UserRibbonService,代码如例 12-14 所示。

【例 12-14】 创建类 UserRibbonService 的代码示例。

```java
package com.bookcode.service;
import com.bookcode.pojo.User;
```

```java
import com.netflix.hystrix.contrib.javanica.annotation.HystrixCommand;
import org.springframework.beans.factory.annotation.Autowired;
import org.springframework.stereotype.Service;
import org.springframework.web.bind.annotation.RequestParam;
import org.springframework.web.client.RestTemplate;
@Service
public class UserRibbonService {
    @Autowired
    RestTemplate restTemplate;
    @HystrixCommand(fallbackMethod = "error")
    public User findUserById(@RequestParam("id") Long id){
        return restTemplate.getForObject("http://service-user/getUserById?id=" + id, User.class);
    }
    public String error(String name){
        String desc = "sorry,访问service-user服务出错,根据name获取user失败,name=" + name;
        return desc;
    }
    public String Hello(){
        return restTemplate.getForObject("http://service-user/hi",String.class);
    }
}
```

12.3.4 创建类 UserController

在包 com.bookcode 中创建子包 controller,并在包 com.bookcode.controller 中创建类 UserController,代码如例 12-15 所示。

【例 12-15】 创建类 UserController 的代码示例。

```java
package com.bookcode.controller;
import com.bookcode.pojo.User;
import com.bookcode.service.UserRibbonService;
import org.springframework.beans.factory.annotation.Autowired;
import org.springframework.web.bind.annotation.*;
@RestController
public class UserController {
    @Autowired
    private UserRibbonService userRibbonService;
    @RequestMapping(value = "/findUserById",method = RequestMethod.GET)
    public String findUserById(@RequestParam Long id){
        User user = userRibbonService.findUserById(id);
        String strUser = "通过feign访问SERVICE-USER服务找不到user";
        if(user != null)        strUser = user.toString();
```

```
        return strUser;
    }
    @GetMapping("/hello")
    public String hello()
    {
        return userRibbonService.Hello();
    }
}
```

12.3.5　创建配置文件 application.yml

在目录 src/main/resources 下,创建配置文件 application.yml,代码如例 12-16 所示。

【例 12-16】　创建的配置文件 application.yml 代码示例。

```
eureka:
  client:
    serviceUrl:
      defaultZone: http://localhost:8761/eureka/
server:
  port: 8093
spring:
  application:
    name: user-ribbon
```

12.3.6　修改入口类

修改入口类,修改后的代码如例 12-17 所示。

【例 12-17】　修改后的入口类代码示例。

```
package com.bookcode;
import org.springframework.boot.SpringApplication;
import org.springframework.boot.autoconfigure.SpringBootApplication;
import org.springframework.cloud.client.discovery.EnableDiscoveryClient;
import org.springframework.cloud.client.loadbalancer.LoadBalanced;
import org.springframework.context.annotation.Bean;
import org.springframework.web.client.RestTemplate;
@EnableDiscoveryClient
@SpringBootApplication
public class DemoApplication {
```

```
@Bean
@LoadBalanced
RestTemplate restTemplate(){
    return new RestTemplate();
}
public static void main(String[] args) {
    SpringApplication.run(DemoApplication.class, args);
}
}
```

12.3.7 运行程序

在12.2节运行程序的基础上，运行本节 case-user-ribbon 程序。在浏览器中输入 localhost:8093/findUserById?id=4，结果如图12-5所示。在浏览器中输入 localhost:8093/hello，结果如图12-6所示。

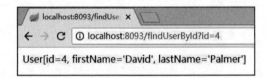

图 12-5　在浏览器中输入 localhost:8093/findUserById?id=4 的结果

图 12-6　在浏览器中输入 localhost:8093/hello 的结果

12.4　实现服务消费端 case-service

视频讲解

12.4.1　创建项目并添加依赖

用 IDEA 创建完项目 case-service 之后，确保在文件 pom.xml 的<dependencies>和</dependencies>之间添加了 Eureka Client、Web、Openfeign 依赖，代码如例12-18所示。

【例12-18】　添加 Eureka Client、Web、Openfeign 依赖的代码示例。

```
<dependency>
    <groupId>org.springframework.cloud</groupId>
    <artifactId>spring-cloud-starter-netflix-eureka-client</artifactId>
</dependency>
<dependency>
```

```xml
            <groupId>org.springframework.boot</groupId>
            <artifactId>spring-boot-starter-web</artifactId>
</dependency>
<dependency>
            <groupId>org.springframework.cloud</groupId>
            <artifactId>spring-cloud-openfeign-core</artifactId>
</dependency>
<dependency>
            <groupId>org.springframework.cloud</groupId>
            <artifactId>spring-cloud-starter-openfeign</artifactId>
</dependency>
```

12.4.2 创建类 User

在包 com.bookcode 中创建 pojo 子包,并在包 com.bookcode.pojo 中创建类 User,代码如例 12-13 所示。

12.4.3 创建接口 UserFeignService

在包 com.bookcode 中创建 service 子包,并在包 com.bookcode.service 中创建接口 UserFeignService,代码如例 12-19 所示。

【例 12-19】 创建接口 UserFeignService 的代码示例。

```java
package com.bookcode.service;
import org.springframework.cloud.openfeign.FeignClient;
import org.springframework.stereotype.Component;
import org.springframework.web.bind.annotation.RequestMapping;
import org.springframework.web.bind.annotation.RequestMethod;
import org.springframework.web.bind.annotation.RequestParam;
@FeignClient(value = "user-ribbon")
@Component
public interface UserFeignService {
    @RequestMapping(value = "findUserById",method = RequestMethod.GET)
    String findByUserId(@RequestParam("id") Long id);
    @RequestMapping(value = "hello",method = RequestMethod.GET)
    String hello();
}
```

12.4.4 创建类 UserController

在包 com.bookcode 中创建子包 controller,并在包 com.bookcode.controller 中

创建类 UserController，代码如例 12-20 所示。

【例 12-20】 创建类 UserController 的代码示例。

```
package com.bookcode.controller;
import com.bookcode.service.UserFeignService;
import org.springframework.beans.factory.annotation.Autowired;
import org.springframework.web.bind.annotation.*;
@RestController
public class UserController {
    @Autowired
    private UserFeignService userFeignService;
    @RequestMapping(value = "/findUserById",method = RequestMethod.GET)
    public String findUserById(@RequestParam Long id){
        return userFeignService.findByUserId(id);
    }
    @GetMapping("/greet")
    public String greet(){
        return userFeignService.hello();
    }
}
```

12.4.5 修改配置文件 application.properties

修改目录 src/main/resources 下配置文件 application.properties，修改后的代码如例 12-21 所示。

【例 12-21】 修改后的配置文件 application.properties 代码示例。

```
server.port = 9000
spring.application.name = user-feign
eureka.client.service-url.defaultZone = http://localhost:8761/eureka/
```

12.4.6 修改入口类

修改入口类，修改后的代码如例 12-22 所示。

【例 12-22】 修改后的入口类代码示例。

```
package com.bookcode;
import org.springframework.boot.SpringApplication;
import org.springframework.boot.autoconfigure.SpringBootApplication;
```

```
import org.springframework.cloud.client.discovery.EnableDiscoveryClient;
import org.springframework.cloud.openfeign.EnableFeignClients;
@SpringBootApplication
@EnableDiscoveryClient
@EnableFeignClients
public class DemoApplication {
    public static void main(String[] args) {
        SpringApplication.run(DemoApplication.class, args);
    }
}
```

12.4.7 运行程序

在12.3节运行程序的基础上,运行本节case-service程序。在浏览器中输入localhost:9000/findUserById?id=6,结果如图12-7所示。在浏览器中输入localhost:9000/greet,结果如图12-8所示。

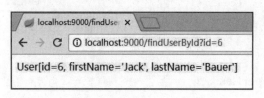

图12-7 在浏览器中输入localhost:9000/findUserById?id=6的结果

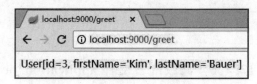

图12-8 在浏览器中输入localhost:9000/greet的结果

12.5 实现路由网关case-zuul

视频讲解

12.5.1 创建项目并添加依赖

用IDEA创建完项目case-zuul之后,确保在文件pom.xml的<dependencies>和</dependencies>之间添加了Zuul、Web、Eureka Client、Security、Actuator依赖,代码如例12-23所示。

【例12-23】 添加Zuul、Web、Eureka Client、Security、Actuator依赖的代码示例。

```
<dependency>
    <groupId>org.springframework.cloud</groupId>
    <artifactId>spring-cloud-starter-netflix-zuul</artifactId>
```

```xml
</dependency>
<dependency>
        <groupId>org.springframework.boot</groupId>
        <artifactId>spring-boot-starter-web</artifactId>
</dependency>
<dependency>
        <groupId>org.springframework.cloud</groupId>
        <artifactId>spring-cloud-starter-netflix-eureka-client</artifactId>
</dependency>
<dependency>
        <groupId>org.springframework.boot</groupId>
        <artifactId>spring-boot-starter-security</artifactId>
</dependency>
<dependency>
        <groupId>org.springframework.boot</groupId>
        <artifactId>spring-boot-starter-actuator</artifactId>
</dependency>
```

12.5.2 创建配置文件 application.yml

在目录 src/main/resources 下，创建配置文件 application.yml，代码如例 12-24 所示。

【例 12-24】 创建的配置文件 application.yml 代码示例。

```yaml
eureka:
  client:
    serviceUrl:
      defaultZone: http://localhost:8761/eureka/
server:
  port: 9003
spring:
  application:
    name: service-zuul
  security:
    basic:
      enabled: true              # 开启基于 HTTP basic 的认证
      user:
        name: zs                 # 配置登录账号
        password: zs             # 配置登录密码
zuul:
  routes:
    api-a:
      path: /api-a/**
      serviceId: service-user
```

```
        api-b:
          path: /api-b/**
          serviceId: user-ribbon
        api-c:
          path: /api-c/**
          serviceId: user-feign
```

12.5.3 修改入口类

修改入口类,修改后的代码如例 12-25 所示。

【例 12-25】 修改后的入口类代码示例。

```
package com.bookcode;
import org.springframework.boot.SpringApplication;
import org.springframework.boot.autoconfigure.SpringBootApplication;
import org.springframework.cloud.netflix.zuul.EnableZuulProxy;
@EnableZuulProxy
@SpringBootApplication
public class DemoApplication {
    public static void main(String[] args) {
        SpringApplication.run(DemoApplication.class, args);
    }
}
```

12.5.4 运行程序

在 12.4 节运行程序的基础上,运行本节 case-zuul 程序。在浏览器中输入 localhost:9003/api-a/getUserById?id=7,页面跳转到 localhost:9003/login,结果如图 12-9 所示。输入正确的 User 和 Password 后,跳回到 localhost:9003/api-a/getUserById?id=7;结果如图 12-10 所示。在浏览器中输入 localhost:9003/api-a/hi,结果如图 12-11 所示。在浏览器中输入 localhost:9003/api-b/findUserById?id=8,结果如图 12-12 所示。在浏览器中输入 localhost:9003/api-c/findUserById?id=8,结果如

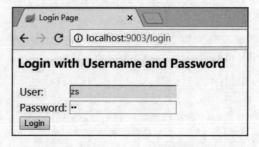

图 12-9　在浏览器中输入 localhost:9003/api-a/getUserById?id=7,页面跳转到 http://localhost:9003/login 的结果

图 12-13 所示。在浏览器中输入 localhost:9003/actuator/health，结果如图 12-14 所示。在浏览器中输入 localhost:8761，结果如图 12-15 所示。

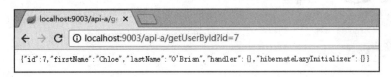

图 12-10　跳回到 localhost:9003/api-a/getUserById? id=7 的结果

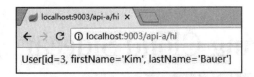

图 12-11　在浏览器中输入 localhost:9003/api-a/hi 的结果

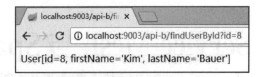

图 12-12　在浏览器中输入 localhost:9003/api-b/findUserById? id=8 的结果

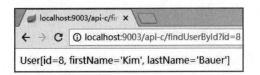

图 12-13　在浏览器中输入 http://localhost:9003/api-c/findUserById? id=8 的结果

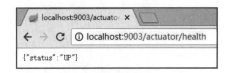

图 12-14　在浏览器中输入 localhost:9003/actuator/health 的结果

Instances currently registered with Eureka

Application	AMIs	Availability Zones	Status
SERVICE-USER	n/a (2)	(2)	UP (2) - LAPTOP-H5IO2QIN:service-user:8091
SERVICE-ZUUL	n/a (1)	(1)	UP (1) - LAPTOP-H5IO2QIN:service-zuul:9003
USER-FEIGN	n/a (1)	(1)	UP (1) - LAPTOP-H5IO2QIN:user-feign:9000
USER-RIBBON	n/a (2)	(2)	UP (2) - LAPTOP-H5IO2QIN:user-ribbon:8095

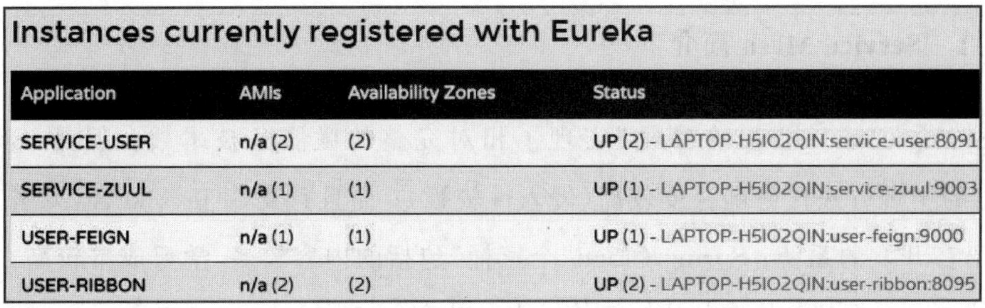

图 12-15　在浏览器中输入 localhost:8761 的结果

习题 12

实验题

请独立完成本章的案例。

第 13 章

Service Mesh 与 Spring Cloud Sidecar

13.1 Service Mesh 概述

13.1.1 Service Mesh 简介

Spring Cloud 通过"全家桶"实现了相对完整的微服务技术栈。但是，Spring Cloud 的实现方式代码耦合度较高（侵入性较强）；而且只支持 Java 语言，无法支持其他语言开发的系统；Spring Cloud"全家桶"包括的内容较多，学习成本较高，并且将系统中已有框架替换成其他框架的成本也较高。

在实施微服务的过程中，除了服务发现、配置中心和鉴权管理等基础功能之外，在服务治理层面也面临了诸多的挑战（如负载均衡、熔断降级、灰度发布、故障切换、跟踪监控等）；这对开发团队提出了非常高的技术要求。在实际应用中，多语言的技术栈和跨语言的服务调用是一种常态，但目前开源社区还没有一套统一的跨语言微服务技术栈，而跨语言调用是微服务概念诞生之初要实现的重要特性之一。

为了解决微服务中的这些问题，Service Mesh（服务网格）概念在 2016 年被提出来。服务网格是一个基础设施层,它的功能在于处理服务间通信,它的职责是实现请求的可靠传递。在实践中,服务网格通常被实现成轻量级的网络代理,与应用程序部署在一起,对应用程序透明。服务网格以基于 Sidecar 模式为主要实现方式,其工作原理如图 13-1 所示。

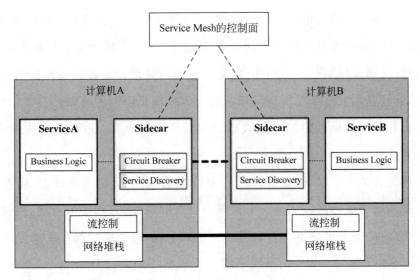

图 13-1　Service Mesh 工作原理

Sidecar(边车)模式允许向应用无侵入地添加多种功能,避免了添加第三方组件时需要向应用添加额外配置代码的工作,此模式也称作"挎斗",主要是因为它类似于三轮摩托车上的挎斗。在此模式中,挎斗附加到父应用程序中,为父应用程序提供支持性功能。挎斗与父应用程序具有相同的生命周期,与父应用程序一起创建和停用。因此,挎斗模式有时也称为搭档模式。

Service Mesh 发展至今也经历了两代。第一代 Service Mesh 的代表是 Linkerd 和 Envoy。Linkerd 基于 Twitter 的 Fingle,使用 Scala 编写,是业界第一个开源的 Service Mesh 方案,在长期的实际生产环境中获得验证。Envoy 底层基于 C++,性能上优于 Linkerd。同时,Envoy 社区成熟度较高,商用稳定版本面世时间也较长。这两个开源实现都是以 Sidecar 为核心,绝大部分关注点都是如何做好代理,并完成一些通用控制面的功能。在容器中大量部署 Sidecar 以后,如何管理和控制这些 Sidecar 是一个不小的挑战。

第二代 Service Mesh 的主要改进集中在提供更加强大的控制面功能,典型代表有 Istio 和 Conduit。Istio 是 Google、IBM 和 Lyft 合作的开源项目,是目前最主流的

Service Mesh 方案,也是事实上的第二代 Service Mesh 标准。在 Istio 中,直接把 Envoy 作为 Sidecar。除了 Sidecar,Istio 中的控制面组件都是使用 Go 语言编写的。

13.1.2 Service Mesh 的特点

Service Mesh 是层次化、规范化、体系化、无侵入的分布式服务治理技术平台。

Service Mesh 分为数据面和控制面两个概念,数据面是指所有数据流动的那个层面,控制面是用来控制这个数据面的,对服务进行处理。对数据面和控制面进行分层,带来的好处是当对复杂系统进行切分时可以获得更清晰的认识。

规范化是指通过标准协议完成数据面和控制面的连接,Sidecar 成为所有互联、互通的约束标准。

体系化包含两个维度,一是指监测的全局考虑,系统化地考虑日志、度量和跟踪这3个监测领域的核心内容。另一个是集中管理的维度,包括服务管理、限流、熔断、安全、灰度在内的服务模块都可以获得体系化的呈现,每个服务都可以被看到。

无侵入是指新增一个服务时不需要用一个 SDK(Software Development Kit,软件开发工具包)去初始化,而是可以通过 Sidecar 进程的方式来解耦式地增加一个服务。

13.1.3 数据面和控制面

数据面的责任是确保请求可靠、安全、及时地从一个微服务传递到另一个微服务,负责提供服务发现、可用性检查、路由、负载平衡、断路器、保护和访问控制、度量监控、日志记录和分布式可追溯性等功能。Linkerd 和 Envoy 可以作为数据面的工具。HAProxy、Træfik 和 NGINX 这样的负载均衡器也因被重新定位为数据面而越来越受欢迎。

控制面将负责管理和监督的所有实例,作为理想的指标收集、监测点,它是服务网格正常运行的必要部分。控制面为现有数据中的所有级别配置服务网,控制面将所有数据面转换为分布式系统。目前有不同类型的服务网格解决方案,例如,Nelson 使用 Envoy 作为其代理 Sidecar 并使用 Hashicorp 堆栈构建强大的控制面(如 Nomad、Consul、Vault 等)。此外,Smartstack 在 HAProxy 或 NGINX 以及其他元素(如 Nerve、Synapse 或 ZooKeeper)周围创建了一个控制面,同时也证明了可以解耦控制面和数据面。Istio 和 Conduit 是目前最流行的两种控制面工具。

13.2 Linkerd 和 Envoy 简介

13.2.1 Linkerd 简介

Linkerd 是一个用于云原生应用的开源可扩展服务网格。Linkerd 用于大公司在运营大型生产系统时发现的问题。最复杂、最令人惊奇和紧急行为的来源通常不是服务本身,而是服务之间的通信。Linkerd 解决了这些问题,Linkerd 不仅仅提供了控制通信机制的功能,还提供了一个抽象层,其工作原理如图 13-2 所示。

通过提供跨服务的统一的仪器仪表和控制层,Linkerd 可以让服务所有者自由选择最适合其服务的语言。通过分离通信机制与应用代码,

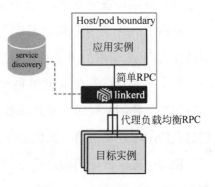

图 13-2　Linkerd 的工作原理

Linkerd 可以对这些机制进行可视化和控制,而无须更改应用本身。

Linkerd 作为独立代理运行,释放其语言和类库要求。应用程序通常通过在已知位置运行 Linkerd 实例来使用 Linkerd,并通过代理调用。Linkerd 应用路由规则与现有服务发现机制来进行通信和目标实例上的负载平衡。

目前,许多组织都在使用 Linkerd 加强软件基础设施的核心。Linkerd 负责跨服务通信中困难而且容易出错的部分工作,包括延迟感知、负载平衡、连接池、传输层安全协议(Transport Layer Security,TLS)、仪器仪表和请求级路由等,让应用代码具备可伸缩性、高性能和弹性。

13.2.2 Envoy 简介

Envoy 作为一个独立的进程与应用程序一起运行。所有的 Envoy 形成一个对应用透明的通信网格,每个应用程序通过本地收发消息而感知不到网络。Envoy 可以与以任何语言开发的应用一起工作,Java、C++、Go、PHP、Python 等都可以基于 Envoy 部署成一个服务网格。在微服务架构中,使用多语言来开发应用的情况越来越普遍,Envoy 填补了这一空白。

Envoy 支持服务发现、健康检查、前端代理、监测与跟踪、动态配置、HTTP/2、gRPC 等功能。

服务发现是微服务架构的重要组成部分。Envoy 支持多种服务发现方法，包括异步 DNS 解析和通过 REST 请求服务、发现服务。

Envoy 含有一个健康检查子系统，它可以对上游服务集群进行主动的健康检查。然后，Envoy 联合服务发现、健康检查信息来确定健康的负载均衡对象。Envoy 作为一个外置健康检查子系统，也支持被动健康检查。

虽然 Envoy 为服务间的通信系统而设计，但也可用于前端。Envoy 提供足够的特性作为 Web 应用的前端代理，包括安全传输层协议、HTTP/1.1、HTTP/2 路由。

Envoy 的目标是使得网络更加透明。然而，无论是网络层还是应用层都有可能出现问题。Envoy 对所有子系统提供可靠的统计能力，可以通过管理端口查看统计信息。Envoy 还支持第三方的分布式跟踪机制。

Envoy 提供分层的动态配置 API，用户可以使用这些 API 构建复杂的集中管理部署。

在 HTTP 模式下，Envoy 支持 HTTP/1.1、HTTP/2，并且支持 HTTP/1.1、HTTP/2 双向代理。这意味着 HTTP/1.1 和 HTTP/2 在客户机和目标服务器的任何组合都可以桥接。建议在服务间的配置使用时，Envoy 之间采用 HTTP/2 来创建持久的网络连接，这样请求和响应可以被多路复用。

gRPC 是一个来自谷歌公司的 RPC 框架，使用 HTTP/2 作为底层的多路传输。HTTP/2 承载的 gRPC 请求和应答，都可以使用 Envoy 的路由和负载均衡能力。所以两个系统非常互补。

13.3　Istio 概述

视频讲解

13.3.1　Istio 简介

使用云平台可以为组织提供较多的好处。然而，采用云平台可能会给 DevOps 团队带来压力。开发人员必须使用微服务以满足应用的可移植性，同时运营商管理了十分庞大的混合云和多云部署。Istio 允许连接、保护、控制和观测服务。

Istio 提供了一个完整的解决方案来满足微服务应用程序的多样化需求。在较高的层次上，Istio 有助于降低部署的复杂性，并减轻开发团队的压力。它是一个完全开源的服务网格，可以透明地分层到现有的分布式应用程序上。它也是一个平台，可以将它集成到日志记录平台、遥测或策略系统的 API。

Istio 提供一种简单的方式来为已部署的服务建立网络。想要让服务支持 Istio，只需要在环境中部署一个特殊的 Sidecar 代理。可以使用 Istio 控制面来配置功能和管理代理，拦截微服务之间的所有网络通信。

Istio 用于实现可扩展性，满足各种部署需求。支持 HTTP、gRPC、WebSocket 和 TCP 流量的自动负载均衡。Istio 通过丰富的路由规则、重试、故障转移和故障注入，可以对流量行为进行细粒度控制。Istio 可插入的策略层和配置 API 支持访问控制、速率限制和配额。Istio 对出入集群的入口和出口中所有流量进行自动度量指标、日志记录和跟踪。Istio 通过强大的基于身份的验证和授权，在集群中实现安全的服务间通信。

13.3.2　Istio 核心功能

Istio 在服务网络中提供了流量管理、安全、监测跟踪、支持多平台、集成、定制和规范化等关键功能。

通过简单的规则配置和流量路由，可以控制服务之间的流量和 API 调用。Istio 简化了断路器、超时和重试等服务级别属性的配置，并且可以轻松地设置 A/B 测试、金丝雀部署和分阶段部署基于百分比的流量分割等重要任务。通过更好地了解流量和开箱即用的故障恢复功能，可以在问题出现之前先发现问题，使调用更可靠，使网络更强大。

Istio 的安全功能使开发人员可以专注于应用程序级别的安全。Istio 提供底层安全通信信道，并大规模地管理服务通信的认证、授权和加密。通过使用 Istio，服务通信在默认情况下是安全的。

Istio 强大的跟踪、监控和日志记录可让用户深入地了解服务网格部署。通过 Istio 的监控功能，可以了解服务性能如何影响上游和下游的功能，而其自定义仪表板可以提供对所有服务性能的可视化。

Istio 独立于平台，用于运行在各种环境中，包括跨云、内部部署、Kubernetes、

Mesos 等。虽然 Istio 与平台无关,但将其与 Kubernetes(或基础架构)网络策略结合使用,其优势会更大。

策略执行组件可以扩展和定制,以便与现有的日志、监控、审计等方案集成。

13.3.3 Istio 架构

Istio 服务网格逻辑上分为数据面和控制面。数据面由一组以 Sidecar 方式部署的代理(proxy)组成。这些代理可以调节和控制微服务以及 Mixer 之间的所有网络通信。控制面负责管理和配置代理来实现路由。此外,控制面配置 Mixer 以实施策略(policy)和收集遥测数据(telemetry)。图 13-3 显示了 Istio 中构成每个面板的不同组件,即 Istio 的架构情况。

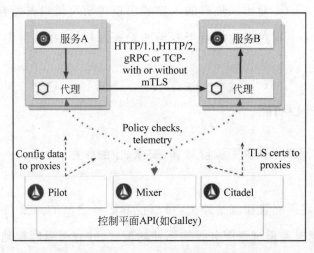

图 13-3　Istio 架构图

Istio 使用 Envoy 代理的扩展版本调解服务网格中所有服务的入站和出站流量。Envoy 的许多内置功能被 Istio 发扬光大。Envoy 被部署为 Sidecar(即图 13-3 中代理),和对应服务在同一个 Kubernetes Pod 中。这允许 Istio 将大量关于流量行为的信号作为属性提取出来,而这些属性又可以在 Mixer 中用于执行策略决策,并发送给监控系统,以提供整个网格行为的信息。

Mixer 是一个独立于平台的组件,负责在服务网格上执行访问控制和使用策略,并从 Envoy 代理和其他服务中收集遥测数据。代理提取请求级属性,发送到 Mixer 进行评估。Mixer 中包括一个灵活的插件模型,使其能够接入到各种主机环境和基础设施后端,从这些细节中抽象出 Envoy 代理和 Istio 管理的服务。

Pilot 为 Envoy Sidecar 提供服务发现功能，为智能路由（例如 A/B 测试、金丝雀部署等）和弹性（超时、重试、熔断器等）提供流量管理功能。它将控制流量行为的高级路由规则转换为特定于 Envoy 的配置（config），并在运行时将它们传播到 Sidecar。

Pilot 将平台特定的服务发现机制抽象化并将其合成为符合 Envoy 数据面 API 的任何 Sidecar 都可以使用的标准格式。这种松散耦合使得 Istio 能够在多种环境下运行（如 Kubernetes），同时保持用于流量管理的相同操作界面。

Citadel 通过内置身份和凭证管理可以提供强大的服务间和最终用户身份验证。可用于升级服务网格中未加密的流量，并为运维人员提供基于服务标识而不是网络控制的强制执行策略。

Galley 代表其他的 Istio 控制面组件，用于验证用户编写的 Istio API 配置。随着时间的推移，Galley 将接管 Istio 获取配置、处理和分配组件的顶级责任。它负责将其他的 Istio 组件与从底层平台（如 Kubernetes）获取用户配置的细节中分离开来。

13.3.4　Istio 应用的模拟

在浏览器中输入 https://www.katacoda.com/courses/istio/deploy-istio-on-kubernetes，可以进行 Istio 应用的模拟操作，如图 13-4 所示。

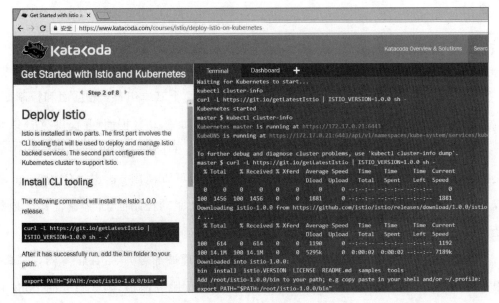

图 13-4　Istio 应用的模拟操作

13.4 Conduit 概述

13.4.1 Conduit 简介

Conduit 是一个面向 Kubernetes 系统的超轻量 Service Mesh。它对运行在 Kubernetes 系统中的服务间通信进行透明管理，让服务变得更加安全可靠和可监控。

Conduit 被设计成可以无缝地融入现有的 Kubernetes 系统。该设计有两个重要特征。

首先，Conduit CLI 设计成尽可能地与 kubectl 一起使用。这是为了在服务网格和 Orchestrator 之间提供一个清晰的工作分工，且能更加容易地适配 Conduit 到已有的 Kubernetes 工作流。

其次，Conduit 在 Kubernetes 中的核心名词是 Deployment，而不是 Service。Deployment 要求单个 Pod 最多是一个 Deployment 的一部分。通过基于 Deployment 而不是 Service 来构建，流量与 Pod 间的映射总是清晰的。

这两个设计特性可以较好地进行组合。例如，Conduit Inject 可用于一个运行的 Deployment，因为当它更新 Deployment 时，Kubernetes 会回滚 Pods 以包含数据面代理。

13.4.2 Conduit 架构

Conduit 服务网格部署到 Kubernetes 集群时有两个基本组件：数据面和控制面。数据面承载服务实例间的实际应用请求流量，而控制面驱动数据面，并提供 API 以修改其行为（还有访问聚合指标）。Conduit CLI 和 Web UI 使用 API，并提供适用于人类的人体工学控制。为了支持 Conduit 的人机交互，可以使用 Conduit CLI、Web UI 或相关工具（如 kubectl）。CLI 和 Web UI 通过 API 驱动控制面，而控制面相应地驱动数据面的行为。

Conduit 的数据面由轻量级的代理组成，这些代理被部署为 Sidecar 容器，与每个服务代码的实例结合在一起。要增加服务到 Conduit 服务网格，则该服务的 Pods 必须重新部署，以便在每个 Pod 中包含一个数据面。

这些代理透明地拦截进出每个 Pod 的通信,并增加重试、超时、仪表和加密等特性,甚至根据相关策略来允许和禁止请求。这些代理并未设计成通过手动方式配置,它们的行为是由控制面驱动的。

Conduit 控制面是一系列服务,运行在专用的 Kubernetes 命名空间(默认是 Conduit)。这些服务完成聚合遥测数据、提供面向用户的 API、为数据面代理提供控制数据等任务。Conduit 控制面还为构建自定义功能提供便利。

Conduit 0.5 成为 Conduit 最后一个主要发布版本,并且 Conduit 逐步进入 Linkerd 项目,成为 Linkerd 2.0 的基础。

Linkerd 2.0 版本为 Linkerd 带来了性能、资源消耗和易用性等方面的显著改善。它还将项目从集群范围的服务网格转换为可组合的服务 Sidecar,旨在为开发人员和服务所有者提供云原生环境中所需的关键工具。Linkerd 2.0 的服务 Sidecar 设计使开发人员和服务所有者能够在他们的服务上运行 Linkerd,提供自动的可观察性、可靠性和运行时诊断,而无须更改配置或代码。通过提供轻量级的增量路径来获得平台范围的遥测、安全性和可靠性的传统服务网格功能,服务 Sidecar 方法还降低了平台所有者和系统架构师的风险。

13.5 国内 Service Mesh 实践简介

13.5.1 SOFAMesh 简介

蚂蚁金服自主研发的分布式中间件(Scalable Open Financial Architecture, SOFA)推出了开源产品 SOFAMesh。SOFAMesh 是基于 Istio 改进和扩展而来的 Service Mesh 大规模实践方案。在继承 Istio 强大功能和丰富特性的基础上,为满足大规模部署下的性能要求和应对实践中的实际情况,对 Istio 进行了一些改进。SOFAMesh 采用 Go 语言编写的边车 MOSN(Modular Observable Smart Netstub)取代 Envoy;合并 Mixer 到数据面以解决性能瓶颈;增强 Pilot 以实现更灵活的服务发现机制;增加对 SOFA RPC、Dubbo 的支持。

13.5.2 Dubbo Mesh 简介

Dubbo 原先是为了 Java 语言而准备的,没有考虑到跨语言的问题。这意味

Node.js、Python、Go 等语言要想无缝地使用 Dubbo 服务，要么需要借助于各自语言的 Dubbo 客户端，如 node-dubbo-client；要么需要借助于服务网格的解决方案，让 Dubbo 自己提供跨语言的解决方案来屏蔽不同语言的处理细节。Dubbo 生态系中的跨语言服务网格解决方案被命名为 Dubbo Mesh。其工作原理如图 13-5 所示。

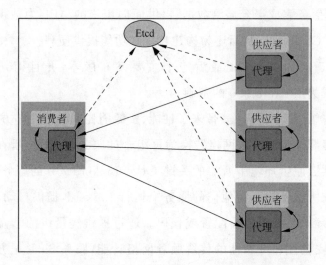

图 13-5　Dubbo Mesh 工作原理

在 Dubbo 原有的生态系下，只有服务提供者、消费者、注册中心等概念。Dubbo Mesh 生态系下可以为每个服务（每个提供者、消费者的实例）启动一个代理（Agent）；服务间不再进行直接的通信，而是经由各自的 Agent 完成交互，并且服务的注册和发现工作也由 Agent 完成。

13.5.3　华为服务网格简介

华为云 Istio 服务网格产品与云容器引擎（Cloud Container Engine，CCE）深度整合，提供非侵入、智能流量治理的应用全生命周期管理方案，并在易用性、可靠性、可视化等方面进行了一系列增强，为客户提供开箱即用的上手体验。华为云 Istio 服务网格内置了金丝雀、A/B 测试等多种灰度发布策略。用户还可自定义配置，实现更多的复杂流量策略。

华为云 Istio 服务网格提供了可视化的流量监控、异常响应、超长响应时延、流量状态信息拓扑等；同时，结合华为云 AOM（Application Operations Management）/APM（Application Performance Management）服务，提供了详细的微服务级流量监控、异常响应流量报告以及调用链信息，实现更加快速的问题定位。

华为云 Istio 服务网格支持根据微服务的流量协议,提供策略化、场景化的网络连接和安全策略管理能力。华为云 Istio 服务网格支持基于应用拓扑对服务配置负载均衡、熔断容错等治理规则,并提供实时的、可视化的微服务流量管理。通过使用华为云 Istio 服务网格,应用无须进行任何改造,即可实现动态的智能路由和弹性流量管理。

13.5.4　京东服务网格简介

京东的服务网格叫 Container Mesh,可分为 3 个部分。其中,基础设施层主要解决网格服务的定义、装配、环境准备和应用发布。数据面主要完成通信协议的编解码、序列化以及相关的服务治理的内容,如熔断、错误注入、流量拆分等。控制面接收用户的配置信息,然后将配置信息传递给数据面;数据面按照配置信息来完成具体的动作。例如,当用户要求服务快速失败时,可以由控制面下达"错误注入"相关的配置指令使得数据面迅速返回错误,从而达到快速失败的效果。

数据面的 Envoy 将同属于一个 Pod 的本地 Service 的流量接管(如通过 iptables 机制)过来,然后 Envoy 询问控制面(Jpilot)目标服务的实例信息(ip/port),这样 Envoy 就知道如何访问目标服务了。控制面 Jpilot 从 JSF Registry 获取服务列表,等待 Envoy 的查询;Envoy 通过与 JSF Registry 的通信,完成服务注册和心跳检测。

13.5.5　新浪微博 Weibo Mesh 简介

以微博应对突发热点事件带来的峰值流量冲击为例,为了确保首页信息流业务的稳定性,需要自动扩缩容系统。开发一套稳定可用的自动扩缩容系统并非一朝一夕之事。随着微博的不断发展,不断涌现出新的业务线,比如热门微博和热搜,也同样面临着突发热点事件带来的流量冲击压力。如何能够把信息流业务研发的自动扩缩容系统推广到各个业务线,也是一个比较棘手的问题。由于新浪信息流业务的后端主要采用了 Java 语言实现,而热门微博和热搜主要采用的是 PHP 语言,无法直接接入自动扩缩容系统。Weibo Mesh 发展之初的首要目的就是想让微博内部的 Motan 服务化框架能够支持 PHP 应用与 Java 应用之间的调用,因而开发了 Motan-go Agent,并在此基础上演变成 Weibo Mesh。Weibo Mesh 支持多种语言之间的服务化调用,有助于统一内部业务不同语言所采用的服务化框架,达到统一技术体系的目的。

13.5.6 云帮 Rainbond 服务网格简介

针对微服务架构的支持,除了兼容已有的微服务架构以外,Rainbond 原生提供了 Service Mesh 架构的支持,对单体应用、新老应用实现规模化整合提供了标准的、完整的功能支持。通过对跨语言、跨协议、代码无侵入的 Service Mesh 微服务架构的原生支持,传统应用直接变成微服务架构。Rainbond 支持常见微服务架构 Spring Cloud、Dubbo 等,还可以通过插件扩展架构能力及治理功能。

Rainbond 原生提供全量的 Service Mesh 治理功能方案;同时提供了插件化的扩展策略,用户除了使用默认方案以外也可以自定义插件实现。Rainbond 与 Istio 的实现有共同点,也有不同点。相同点是二者都实现了全局控制层,不同点是 Istio 需要依赖 Kubernetes 等平台工作。微服务架构的支持需要从底层存储与通信到上层的应用层配置方面进行全盘考虑,大型的微服务架构离不开自动化管理应用的 PaaS 平台。Rainbond 从硬件层、通信层、平台层实现不同的控制逻辑,既兼容已有的微服务架构,同时又提供了完整的 Service Mesh 微服务架构实践,其工作原理如图 13-6 所示。包容的架构形式让已有的应用服务化变得可行。

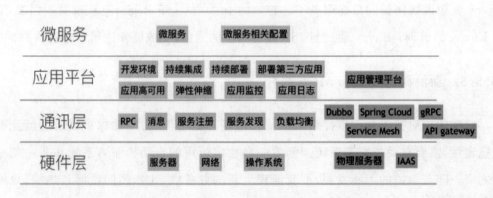

图 13-6　Rainbond 微服务架构工作原理图

13.6　Spring Cloud Sidecar 的应用

视频讲解

非 JVM 微服务可操作 Eureka 的 REST 端点,从而实现服务注册与发现。事实上,也可使用 Sidecar 更方便地整合非 JVM 微服务。Spring Cloud

Netflix Sidecar 的灵感来自于 Netflix Prana，它包括了一个简单的 HTTP API 来获取指定服务所有实例信息（例如，主机和端口）。

13.6.1 创建项目并添加依赖

用 IDEA 创建完项目 sidecarexample 之后，确保在文件 pom.xml 的 < dependencies > 和 </dependencies > 之间添加了 Sidecar、Eureka Client 依赖，代码如例 13-1 所示。

【例 13-1】 添加 Sidecar、Eureka Client 依赖的代码示例。

```
<dependency>
        <groupId>org.springframework.cloud</groupId>
        <artifactId>spring-cloud-netflix-sidecar</artifactId>
</dependency>
<dependency>
        <groupId>org.springframework.cloud</groupId>
        <artifactId>spring-cloud-starter-netflix-eureka-client</artifactId>
</dependency>
```

13.6.2 修改配置文件 application.properties

修改配置文件 application.properties，修改后的代码如例 13-2 所示。

【例 13-2】 修改后的配置文件 application.properties 的代码示例。

```
server.port=8082
spring.application.name=sidecar-server
eureka.client.service-url.defaultZone = http://localhost:8761/eureka/
sidecar.port=8205
sidecar.health-uri=http://localhost:${sidecar.port}/health.json
management.endpoints.web.exposure.include=*
```

13.6.3 修改入口类

修改入口类，修改后的代码如例 13-3 所示。

【例 13-3】 修改后的入口类代码示例。

```
package com.bookcode;
import org.springframework.boot.SpringApplication;
```

```
import org.springframework.boot.autoconfigure.SpringBootApplication;
import org.springframework.cloud.netflix.sidecar.EnableSidecar;
@EnableSidecar
@SpringBootApplication
public class DemoApplication {
    public static void main(String[] args) {
        SpringApplication.run(DemoApplication.class, args);
    }
}
```

13.6.4 创建 node-service.js

创建类 node-service.js，代码如例 13-4 所示。

【例 13-4】 创建类 node-service.js 的代码示例。

```
//nodejs 引入 http、url、path 模块
var http = require('http');
var url = require("url");
var path = require('path');
//创建 server
var server = http.createServer(function(req, res) {
//获得请求的路径
    var pathname = url.parse(req.url).pathname;
    res.writeHead(200, { 'Content-Type' : 'application/json; charset=utf-8' });
//访问 http://localhost:8205/,将会返回{"index":"欢迎来到简单异构系统之 node.js 服务首页"}
    if (pathname === '/') {
        res.end(JSON.stringify({ "index" : "欢迎来到简单异构系统之 node.js 服务首页" }));
    }
//访问 http://localhost:8205/health,将会返回{"status":"UP"}
    else if (pathname === '/health.json') {
        res.end(JSON.stringify({ "status" : "UP" }));
    }
//其他情况返回 404
    else {
        res.end("404");
    }
});
//创建监听,并打印日志
server.listen(8205, function() {
    console.log('开始监听本地端口:8205');
});
```

13.6.5 运行程序

安装 Node.js 的步骤，参考附录 G.1 节。

执行启动命令,代码如例 13-5 所示,启动程序 node-service.js。

【例 13-5】 启动程序 node-service.js 的命令示例。

```
node node-service.js
```

依次启动行 3.2 节中 eureka-server 程序和本节的 sidercarexample 程序。在浏览器中输入 localhost:8205,结果如图 13-7 所示。在浏览器中输入 localhost:8205/health.json,结果如图 13-8 所示。在浏览器中输入 localhost:8082,结果如图 13-9 所示。在浏览器中输入 localhost:8082/ping,结果如图 13-10 所示。在浏览器中输入 localhost:8082/hosts/sidecar-server,结果如图 13-11 所示。在浏览器中输入 localhost:8082/actuator/health,结果如图 13-12 所示。关闭 node-service.js 程序,在浏览器中输入 localhost:8082/actuator/health,结果如图 13-13 所示。

图 13-7　在浏览器中输入 localhost: 8205 的结果

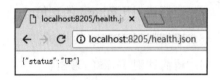

图 13-8　在浏览器中输入 localhost: 8205/health.json 的结果

图 13-9　在浏览器中输入 localhost: 8082 的结果

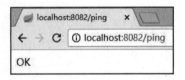

图 13-10　在浏览器中输入 localhost: 8082/ping 的结果

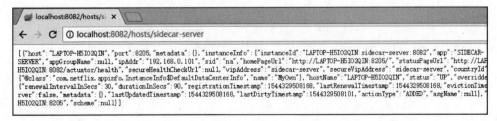

图 13-11　在浏览器中输入 localhost:8082/hosts/sidecar-server 的结果

图 13-12 在浏览器中输入 localhost:8082/actuator/health 的结果(没有关闭 Node.js 程序)

图 13-13 在浏览器中输入 localhost:8082/actuator/health 的结果(关闭 Node.js 服务后)

习题 13

一、问答题

1. 请简述对 Service Mesh 的理解。
2. 请简述对 Linkerd 的理解。
3. 请简述对 Envoy 的理解。
4. 请简述对 Istio 的理解。
5. 请简述对 Sidecar 的理解。

二、实验题

1. 请实现 Istio 的应用。
2. 请实现 Spring Cloud Sidecar 的应用。

附录 A

相关软件的安装和配置

本附录介绍 JDK、Consul、ZooKeeper、Nacos Server 的安装和配置。

A.1 JDK 的安装和配置

Finchley.RELEASE 版的 Spring Cloud 是基于 Spring Boot 2.0.x 构建的。使用 2.0.0 以上版本的 Spring Boot 需要安装 1.8 及以上版本的 JDK。

从 Java 的官网下载安装包,如图 A-1 所示。

图 A-1 从官网下载 JDK 安装包

安装完成后，设置系统变量 JAVA_HOME，如图 A-2 所示。配置好 JAVA_HOME 之后，将％JAVA_HOME％\bin 加入到系统的环境变量 path 中，如图 A-3 所示。

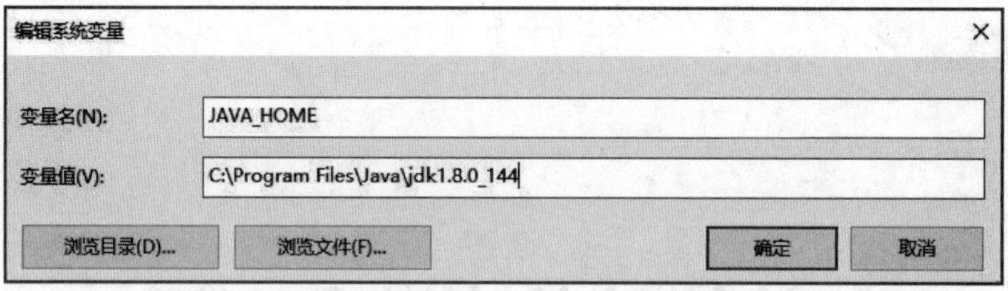

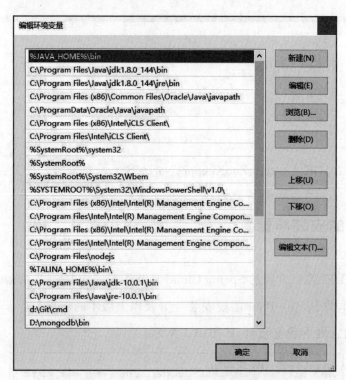

图 A-2　设置系统变量 JAVA_HOME

图 A-3　设置系统变量 path 中 JDK 的路径信息

A.2　Consul 的配置

从 Consul 的官网（https://www.consul.io/）下载压缩包，如图 A-4 所示。

解压后，目录和文件如图 A-5 所示。将 Consul 的路径信息加入到系统的环境变量 path 中，如图 A-6 所示。

附录A 相关软件的安装和配置

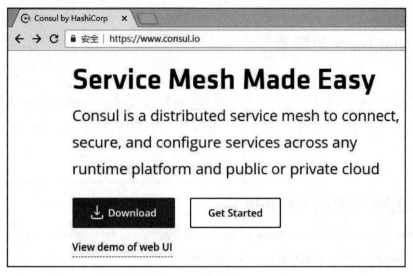

图 A-4　在官网下载 Consul 压缩包

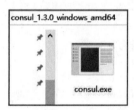

图 A-5　Consul 压缩包解压后的目录和文件

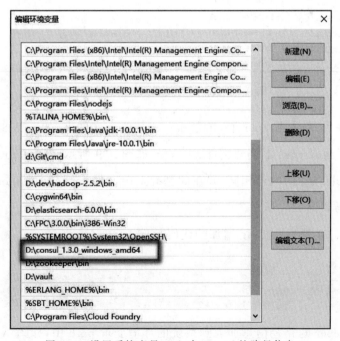

图 A-6　设置系统变量 path 中 Consul 的路径信息

A.3 ZooKeeper 的配置

从 ZooKeeper 的官网(https://archive.apache.org/dist/zookeeper/)或国内镜像网址(如 http://mirrors.hust.edu.cn/apache/zookeeper/)下载压缩包,如图 A-7 所示。

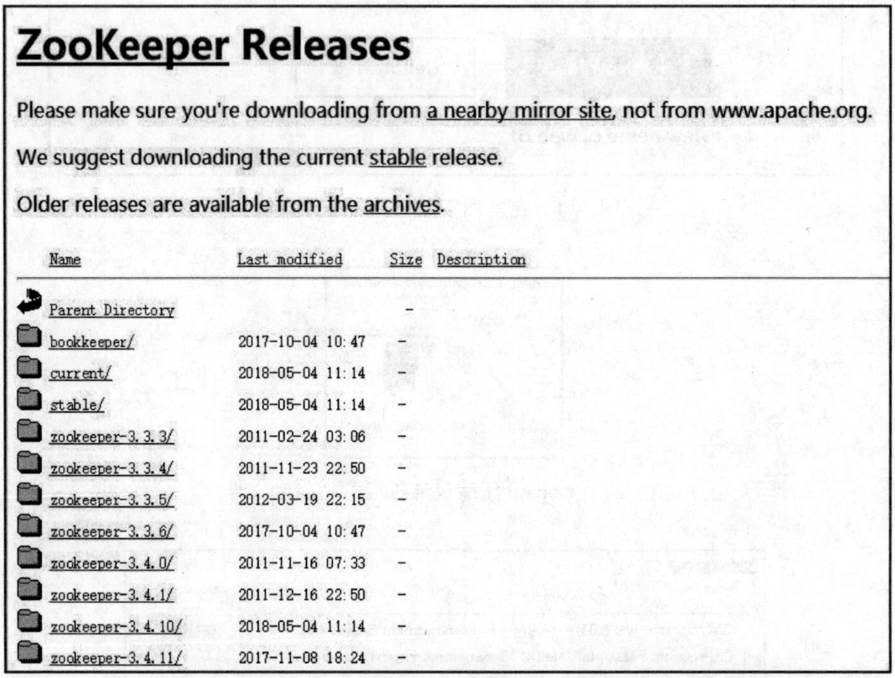

图 A-7 从官网下载 ZooKeeper 压缩包

解压后,在目录 zookeeper 下,建立 data、log 两个文件夹,如图 A-8 所示。在目录 zookeeper 的子目录 conf 下创建文件 zoo.cfg,如图 A-9 所示。文件 zoo.cfg 的内容和文件 zoo_sample.cfg 的内容相同,并在此基础上向文件 zoo.cfg 增加 data 和 log 的文件夹信息,代码如例 A-1 所示。

【例 A-1】 向文件 zoo.cfg 增加的代码示例。

```
dataDir = D:\\zookeeper\\data
dataLogDir = D:\\zookeeper\\log
```

将 ZooKeeper 的路径信息加入到系统的环境变量 path 中,如图 A-10 所示。

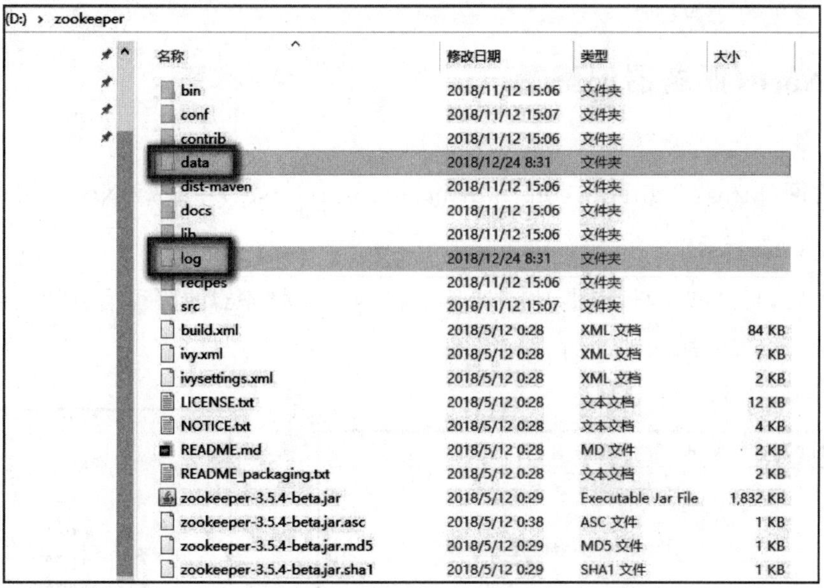

附录A　相关软件的安装和配置

图 A-8　解压压缩包后在目录 zookeeper 下新建 data 和 log 文件夹

图 A-9　conf 目录下创建文件 zoo.cfg

图 A-10　设置系统变量 path 中 ZooKeeper 的路径信息

A.4 Nacos 服务器的配置

在官网(https://github.com/alibaba/nacos/releases)中下载 Nacos 服务器压缩包,如图 A-11 所示。

解压后,目录和文件如图 A-12 所示。将 Nacos 服务器的路径信息加入到系统的环境变量 path 中,如图 A-13 所示。

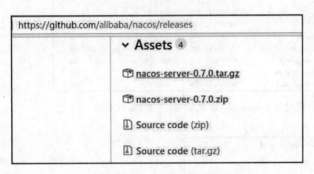

图 A-11　从官网下载 Nacos 服务器压缩包

图 A-12　Nacos 压缩包解压后的目录和文件

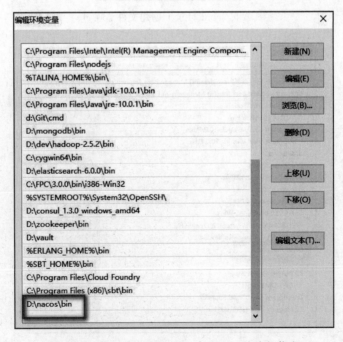

图 A-13　设置系统变量 path 中 Nacos 的路径信息

视频讲解

基于 Feign 实现文件传送

本附录介绍如何基于 Feign 实现文件传送。整个应用分为 3 个微服务项目，分别是 Eureka 服务器（或称为注册中心）项目 mweathereurekaserver、文件接收者项目 feignserver、文件传送者项目 feignclient。

B.1 实现 Eureka 服务器项目 mweathereurekaserver

B.1.1 新建项目并添加依赖

新建 Eureka 服务器项目 mweathereurekaserver，确保在文件 pom.xml 的 <dependencies>和</dependencies>之间添加了 Eureka Server 依赖。

B.1.2 修改入口类与创建配置文件

修改入口类，增加注解@EnableEurekaServer。
在目录 src/main/resources 下创建配置文件 application.yml，并修改其代码。
上述文件的具体代码请参考本书附带的源代码。

B.2 实现文件接收者项目 feignserver

B.2.1 新建项目并添加依赖

新建项目 feignserver,确保在文件 pom.xml 的< dependencies >和</ dependencies >之间添加了 Eureka Client、Openfeign、Web 依赖。

B.2.2 修改入口类和配置文件

在包 com.bookcode 中创建类 UploadController,并修改其代码。

修改在目录 src/main/resources 下的配置文件 application.properties。

上述文件的具体代码请参考本书附带的源代码。

B.3 实现文件传送者项目 feignclient

B.3.1 新建项目并添加依赖

新建项目 feignclient,确保在文件 pom.xml 的< dependencies >和</ dependencies >之间添加了 Eureka Client、Openfeign、Web 依赖。

B.3.2 创建接口、类和修改配置文件

在包 com.bookcode 中创建接口 UploadService 和类 UploadFileController,并修改其代码。

修改在目录 src/main/resources 下的配置文件 application.properties。

上述文件的具体代码请参考本书附带的源代码。

B.4 程序运行结果

在 D 盘创建一个文件 test.txt。

依次运行项目 mweathereurekaserver(端口为 8761)、feignserver(服务名称为

feignserver-upload,端口为 10001)、feignclient(服务名称为 feignclient-upload,端口为 10002)。

在浏览器中输入 localhost:8761,结果如图 B-1 所示,显示服务 feignclient-upload 和服务 feignserver-upload 注册成功。在浏览器中输入 localhost:10001/success,结果如图 B-2 所示。在浏览器中输入 localhost:10002/feignclient/test.txt,浏览器中的结果如图 B-3 所示,同时在控制台中输出"文件 test.txt 传送成功。"。再次在浏览器中输入 localhost:10001/success,浏览器中的结果如图 B-4 所示,同时在控制台中输出"接收到的文件为:test.txt。"。对比图 B-2 和图 B-4,或者根据控制台的输出信息,可以知道文件是否传送成功。

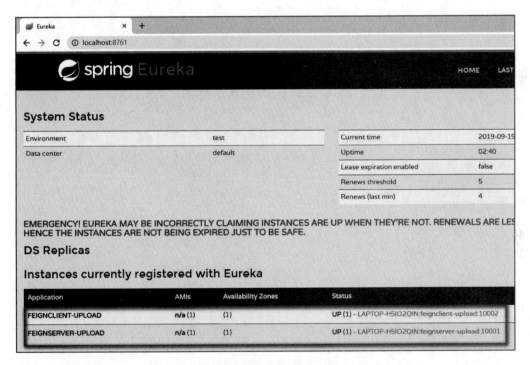

图 B-1 在浏览器中输入 localhost:8761 的结果

图 B-2 在浏览器中输入 localhost:10001/success 的结果

图 B-3 在浏览器中输入 localhost:10002/feignclient/test.txt 后浏览器中的结果

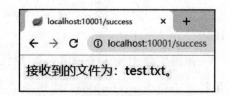

图 B-4 再次在浏览器中输入 localhost:10001/success 后浏览器中的结果

附录 C

视频讲解

基于 Ribbon 实现文件上传

本附录介绍基于 Ribbon 实现文件上传。整个应用分为 3 个微服务项目，分别是 Eureka 服务器项目 mweathereurekaserver、文件上传服务提供者项目 uploadfile、文件上传服务消费者项目 fileuser。

本案例中 Eureka 服务器项目 mweathereurekaserver 的实现方法和附录 B 中 B.1 节的实现方法相同，请参考 B.1 节的说明和本书附带的源代码。

C.1 实现文件上传服务提供者项目 uploadfile

C.1.1 新建项目并添加依赖

新建项目 uploadfile，确保在文件 pom.xml 的< dependencies >和</ dependencies >之间添加了 Eureka Client、Thymeleaf、Lombok、Web 依赖。

C.1.2 创建类

依次在包 com.bookcode 下创建 exception、service、controller 等子包。并在包

com.bookcode.exception中创建类StorageException、StorageFileNotFoundException,在包com.bookcode.service中创建类StorageService,在包com.bookcode.controller中创建类FileUploadController、UploadPictureController,并修改这些类的代码。

C.1.3 新建文件和修改配置文件

在目录src/main/resources/templates下创建文件index.html、welcome.html、uploadForm.html,并修改这些文件的代码。

修改在目录src/main/resources下的配置文件application.properties。

上述文件的具体代码请参考本书附带的源代码。

C.2 实现文件上传服务消费者项目fileuser

C.2.1 新建项目并添加依赖

新建项目fileuser,确保在文件pom.xml的<dependencies>和</dependencies>之间添加了Eureka Client、Web依赖。

C.2.2 创建类、修改配置文件和配置文件

在包com.bookcode中创建类CallFileUpload、FileUploadController,并修改这些类的代码。

修改入口类,增加实现负载均衡的代码。

修改在目录src/main/resources下的配置文件application.properties。

上述文件的具体代码请参考本书附带的源代码。

C.3 程序运行结果

在D盘创建一个文件夹upload-dir。

依次运行项目mweathereurekaserver(端口为8761)、uploadfile(服务名称为file-upload,端口为8093)、fileuser(端口为8099)。

在浏览器中输入 localhost:8093,结果如图 C-1 所示。单击"选择文件"按钮,如图 C-1 所示,选择要上传的文件,再单击 Upload 按钮,结果如图 C-2 所示。在浏览器中输入 localhost:8093/upload,结果如图 C-3 所示。单击图 C-3 中"选择文件"按钮,选择要上传的文件,再单击图 C-3 中"上传"按钮,结果如图 C-2 所示。

图 C-1　在浏览器中输入 localhost:8093 的结果　　　图 C-2　正确上传文件后的结果

图 C-3　在浏览器中输入 localhost:8093/upload 的结果

在浏览器中输入 localhost:8099/uponefile,结果如图 C-4 所示。单击图 C-4 中"选择文件"按钮,选择要上传的文件,再单击 Upload 按钮,结果如图 C-5 所示。在浏览器中输入 localhost:8099/upfiles,结果如图 C-6 所示。单击图 C-6 中"选择文件"按钮,选择要上传的文件,再单击"上传"按钮,结果如图 C-5 所示。

图 C-4　在浏览器中输入 localhost:8099/　　　图 C-5　在图 C-4 或 C-6 的基础上正确
　　　　uponefile 的结果　　　　　　　　　　　　　　上传文件后的结果

图 C-6　在浏览器中输入 localhost:8099/upfiles 的结果

保持原有项目 uploadfile（端口为 8093）中其他所有文件不变，只修改配置文件 application.properties 中设置端口的代码，再启动两个与项目 uploadfile 微服务内容相同的新服务器（端口分别为 8095、8096）。在浏览器中输入 localhost:8099/greet，并不断进行刷新页面操作，会依次循环显示如图 C-7（对应的服务器端口为 8093）、图 C-8（对应的服务器端口为 8095）、图 C-9（对应的服务器端口为 8096）所示的结果。

图 C-7　由端口为 8093 的服务器提供服务的结果

图 C-8　由端口为 8095 的服务器提供服务的结果

图 C-9　由端口为 8096 的服务器提供服务的结果

与本案例类似的代码可以参考网址 https://github.com/JavaCodeMood/spring-cloud-demo/tree/master/file-upload，读者可以对比此网址的代码和本书附带的源代码，加深对 Ribbon 应用开发的认识。

附录 D

视频讲解

简易天气预报系统的实现

本附录介绍简易天气预报系统的实现。整个应用分为 3 个微服务项目，分别是 Eureka 服务器项目 mweathereurekaserver、天气服务提供者项目 weatherbasic、天气服务消费者项目 weatherclient。

本案例中 Eureka 服务器项目 mweathereurekaserver 的实现方法和附录 B 中 B.1 节的实现方法相同，请参考 B.1 节的说明和本书附带的源代码。

D.1 实现天气服务提供者项目 weatherbasic

D.1.1 新建项目并添加依赖

新建项目 weatherbasic，确保在文件 pom.xml 的 < dependencies > 和 </dependencies >之间添加了 Eureka Client、Web、Lombok 等依赖。

D.1.2 创建类、接口并修改配置文件

依次在包 com.bookcode 下创建 vo、service、controller 等子包，并在包 com.bookcode.vo 中创建类 Forecast、Weather、WeatherResponse、Yesterday，在包

com.bookcode.service 中创建接口 WeatherDataService 和 impl 子包，在包 com.bookcode.service.impl 中创建类 WeatherDataServiceImpl，在包 com.bookcode.controller 中创建类 WeatherController，并修改这些类和接口的代码。

修改在目录 src/main/resources 下的配置文件 application.properties。

上述文件的具体代码请参考本书附带的源代码。

D.2 实现天气服务消费者项目 weatherclient

D.2.1 新建项目并添加依赖

新建项目 weatherclient，确保在文件 pom.xml 的< dependencies >和</dependencies >之间添加了 Eureka Client、Thymeleaf、Lombok、Web 依赖。

D.2.2 创建类

依次在包 com.bookcode 下创建 entity、service、controller、utils 等子包，并在包 com.bookcode.entity 中创建类 City、CityList、Forecast、Weather、Yesterday，在包 com.bookcode.service 中创建类 CityDataService、CallWeatherService、WeatherReportService，在包 com.bookcode.controller 中创建类 CallWeatherController、WeatherReportController，在包 com.bookcode.utils 中创建类 XmlBuilder，并修改这些类的代码。

D.2.3 新建文件和修改配置文件

在目录 src/main/resources 下创建文件 citylist.xml，在目录 src/main/resources/templates 下创建文件 report.html，并修改这些文件的代码。

修改在目录 src/main/resources 下的配置文件 application.properties。

上述文件的具体代码请参考本书附带的源代码。

D.3 程序运行结果

依次运行项目 mweathereurekaserver、weatherbasic（服务名称为 microweatherservice，端口为 8762）、weatherclient（端口为 8764）。

在浏览器中输入 localhost:8762/weather/cityId/101190801，结果如图 D-1 所示。在浏览器中输入 localhost:8764/cityId/101190801，结果如图 D-2 所示。对比图 D-1 和图 D-2 可以发现两者数据（天气服务内容）完全相同。在浏览器中输入 localhost:8764/report/cityId/101190801，结果如图 D-3 所示。

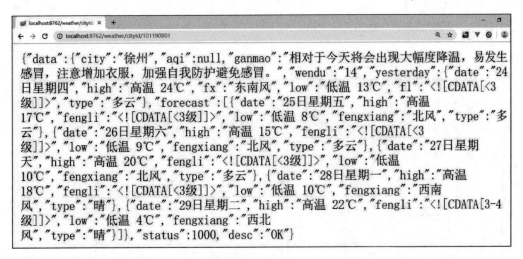

图 D-1　在浏览器中输入 localhost:8762/weather/cityId/101190801 的结果

图 D-2　在浏览器中输入 localhost:8764/cityId/101190801 的结果

读者可以通过天气服务消费者项目 weatherclient 体会到访问远程 Spring Cloud 微服务和实现本地服务的差异。与本系统类似的代码可以参考网址 https://github.com/LuckyShawn/spring-cloud-weather，读者可以对比此网址的代码和本书附带的源代码，加深对 Spring Cloud 微服务开发的认识。

读者还可以对比本案例与用 Spring Boot 实现的天气预报系统（可参考由本书编者编写的《Spring Boot 开发实战-微课视频版》一书或其他资料），体会 Spring Boot 开发和 Spring Cloud 开发的异同。

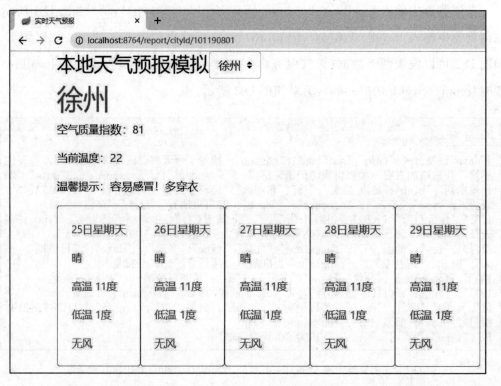

图 D-3　在浏览器中输入 localhost:8764/report/cityId/101190801 的结果

附录 E

视频讲解

Apollo 和 Zuul 的整合开发

Apollo 是携程研发的开源配置管理中心,能够集中管理应用于不同环境、不同集群的配置,配置修改后能够实时推送到应用端,并且具备规范的权限、流程治理等特性。

本附录结合一个案例介绍 Apollo 和 Zuul 的整合开发。整个应用分为 4 个微服务项目,分别是 Eureka 服务器项目 mweathereurekaserver、服务提供者项目 apolloconfig、服务提供者项目 apollouser、服务消费者(即 zuul 路由服务)项目 zuulapollo。

本案例中 Eureka 服务器项目 mweathereurekaserver 的实现方法和附录 B 中 B.1 节的实现方法相同,请参考 B.1 节的说明和本书附带的源代码。

E.1 Apollo 配置中心的准备和启动

E.1.1 Apollo 配置中心的准备

为了让大家更快地了解 Apollo 配置中心(或称为服务器),Apollo 研发者准备了一个 Quick Start 项目,通过该项目能够在几分钟内部署和启动 Apollo 配置中心。先从 Quick Start 的代码库(https://github.com/nobodyiam/apollo-build-scripts)中下载该项目的代码压缩包并进行解压缩。解压缩后的目录和文件如图 E-1 所示。

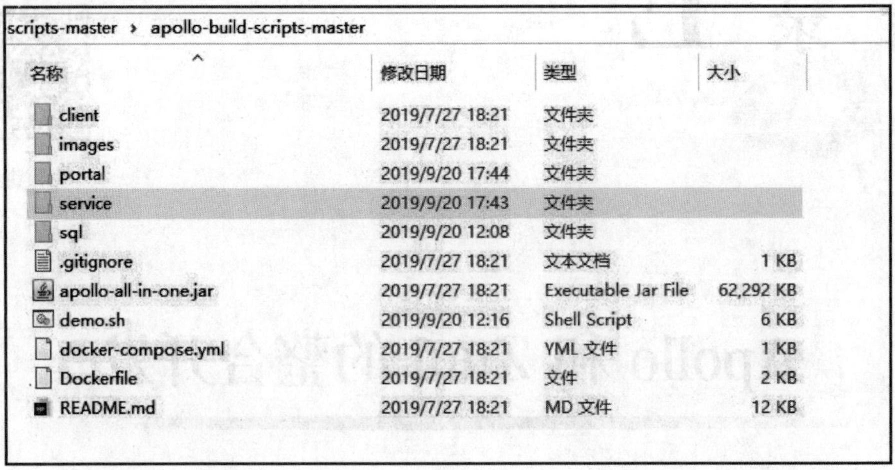

图 E-1　解压缩后 apollo-builds-scripts-master 文件夹内的文件和目录

使用 Apollo 时先要确保安装的 Java 版本在 1.8 以上，安装的 MySQL 版本在 5.6.5 以上。由于 Quick Start 需要用到 Git Bash 环境，需要安装 Git Bash（或者直接使用 IDE 的 Git Bash 环境）。

Apollo 服务端需要两个数据库：ApolloPortalDB 和 ApolloConfigDB。通过 Navicat for MySQL 或 MySQL 原生客户端，导入解压缩包里 sql 目录下的文件 apolloportaldb.sql 和文件 apolloconfigdb.sql。

Apollo 服务端需要知道如何连接到前面创建的两个数据库，所以需要修改文件 demo.sh 中数据库连接信息。将 root 的用户名和密码改为您自己的 MySQL 的 root 用户名和密码。

E.1.2　Apollo 配置中心的启动

在目录 apollo-builds-scripts-master 下启动 Git Bash，执行如例 E-1 所示的命令启动 Apollo 配置中心。

【例 E-1】启动 Apollo 配置中心的命令示例。

```
./demo.sh start
```

启动 Apollo 配置中心的命令、过程和结果如图 E-2 所示。在浏览器中输入 localhost:8070，结果如图 E-3 所示。在图 E-3 中输入正确的 Username（初始值为

apollo)和 Password(初始值为 admin)后,结果如图 E-4 所示,显示已有一个项目默认 SampleApp。SampleApp 项目的基本信息如图 E-5 所示。

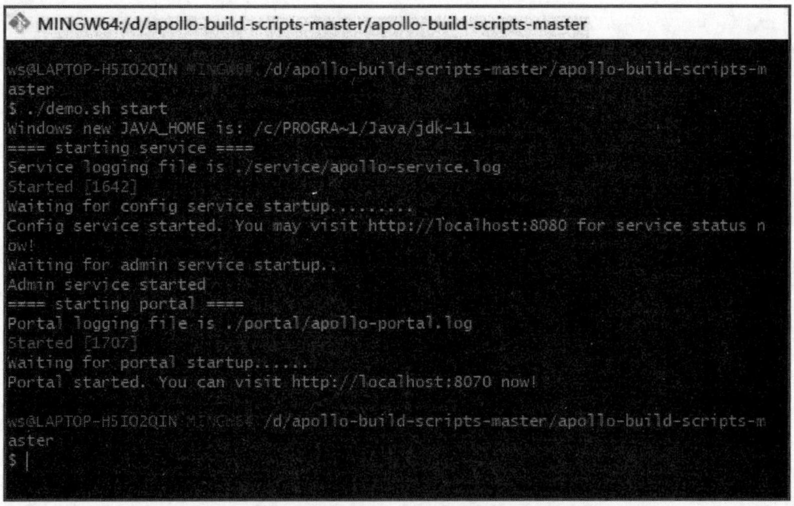

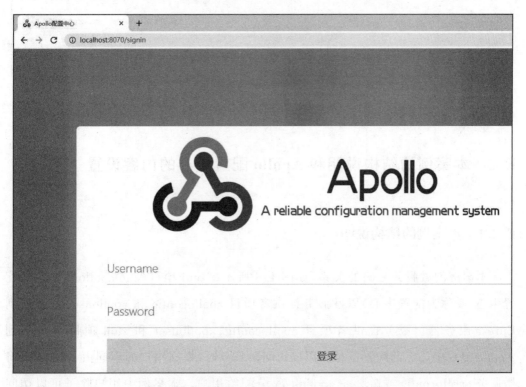

图 E-2　Apollo 配置中心的启动命令、过程和结果

图 E-3　在浏览器中输入 localhost:8070 的结果

图 E-4　在图 E-3 中输入正确 Username 和 Password 后的结果

图 E-5　默认项目 SampleApp 的基本信息

E.2　本案例的结构说明和 Apollo 配置中心的内容设置

E.2.1　本案例的结构说明

本案例的微服务之间的关系，如图 E-6 所示。zuul 项目（zuulapollo）、Apollo 配置中心（或称为配置中心）以及服务提供者项目 apolloconfig 和 apollouser 都要用到 Eureka 服务器。服务提供者项目 apolloconfig、apollouser 和 zuul 项目都要用到 Apollo 配置中心上的配置信息。用户访问微服务时，根据用户的不同由 zuul 项目将微服务 apolloconfig 或微服务 apollouser 分配给用户。本案例中用户除了可以访问 zuul 项目之外还可以直接访问项目 apolloconfig 或 apollouser（正式情况下一般不能直接访问微服务）。为了对比，项目 apolloconfig 或 apollouser 均提供了返回文本内容和返回视图两类接口。

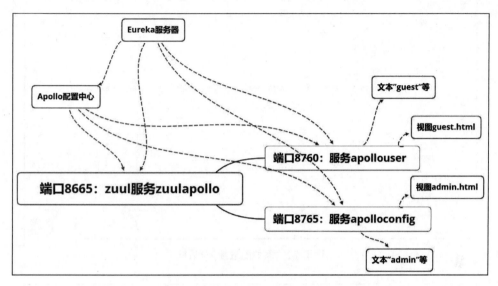

图 E-6　本案例的微服务之间的关系

E.2.2　Apollo 配置中心的内容设置

在 Apollo 配置中心默认项目 SampleApp 中，增加如表 E.1 所示的配置内容。增加 1 条配置信息（以 admin 为例）的方法是单击项目 SampleApp 后，再单击"新增配置"按钮，弹出"添加配置项"对话框，如图 E-7 所示。依次在 Key 文本框和 Value 文本框填写 admin、admin，单击"提交"按钮。

表 E.1　要在 Apollo 配置中心的默认项目 SampleApp 中增加的配置内容

Key	Value
admin	admin
guest	guest
zuul.routes.apollouser.path	/guest/**
zuul.routes.apollouser.serviceId	apollouser

按照同样方法设置表 E.1 中其他配置信息，单击"发布"按钮，结果如图 E-8 所示。

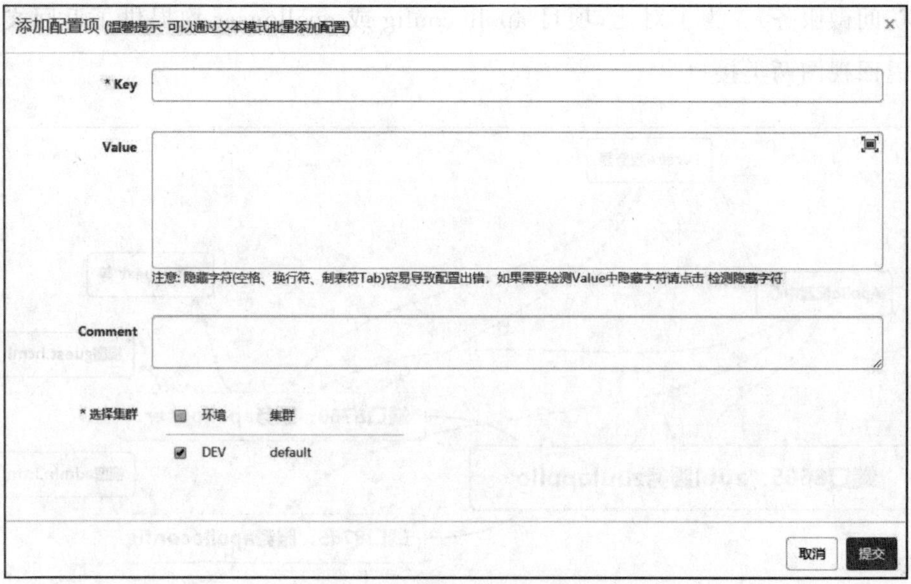

图 E-7 "添加配置项"对话框

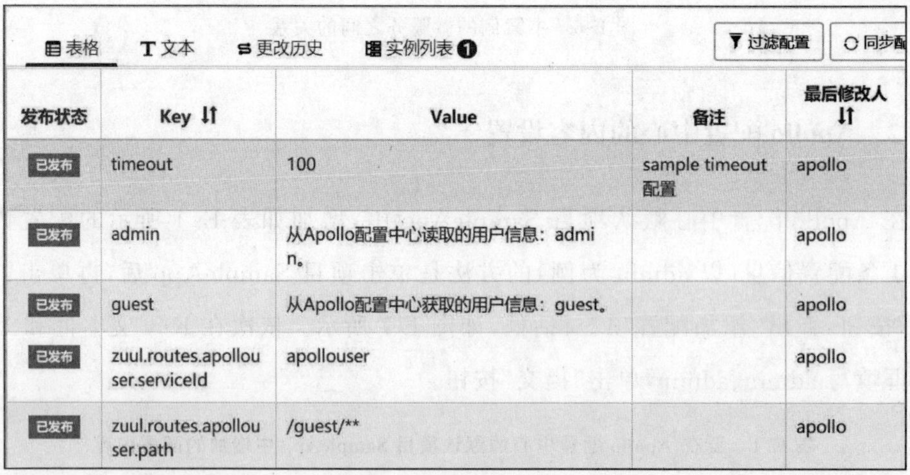

图 E-8 向项目 SampleApp 增加配置内容并进行发布之后的结果

E.3 实现服务提供者项目 apolloconfig

E.3.1 新建项目并添加依赖

新建项目 apolloconfig,确保在文件 pom.xml 的<dependencies>和</dependencies>之间添加了 Eureka Client、Web、Thymeleaf、Apollo Client 依赖。

E.3.2　创建类、文件和修改配置文件

创建类 AppConfig、ACController，并修改这些类的代码。

在目录 src/main/resources/templates 下创建文件 admin.html，并修改其代码。

修改在目录 src/main/resources 下的配置文件 application.properties。

上述文件的具体代码请参考本书附带的源代码。

E.4　实现服务提供者项目 apollouser

E.4.1　新建项目并添加依赖

新建项目 apollouser，确保在文件 pom.xml 的 <dependencies> 和 </dependencies> 之间添加了 Eureka Client、Web、Thymeleaf、Apollo Client 依赖。

E.4.2　创建类、文件和修改配置文件

创建类 AppConfig、ACController，并修改这些类的代码。

在目录 src/main/resources/templates 下创建文件 guest.html，并修改其代码。

修改在目录 src/main/resources 下的配置文件 application.properties。

上述文件的具体代码请参考本书附带的源代码。

E.5　实现 zuul 项目 zuulapollo

E.5.1　新建项目并添加依赖

新建项目 zuulapollo，确保在文件 pom.xml 的 <dependencies> 和 </dependencies> 之间添加了 Eureka Client、Web、Zuul、Apollo Client 依赖。

E.5.2　创建类、修改入口类和配置文件

在包 com.bookcode 中创建类 ZuulPropertiesRefresher，并修改其代码。

修改入口类，增加注解@EnableApolloConfig 和注解@EnableZuulProxy。

修改在目录 src/main/resources 下的配置文件 application.properties。

上述文件的具体代码请参考本书附带的源代码。

E.6 程序运行结果

依次运行项目 mweathereurekaserver（端口为 8761）、apolloconfig（服务名称为 apolloconfig，端口为 8765）、apollouser（服务名称为 apollouser，端口为 8760）、zuulapollo（服务名称为 zuulapollo，端口为 8665）。

E.6.1 apolloconfig 服务运行结果

在浏览器中输入 localhost:8765/userinfo，结果如图 E-9 所示。在浏览器中输入 localhost:8765/admin/userinfo，结果如图 E-10 所示。

图 E-9 在浏览器中输入 localhost:8765/userinfo 的结果

图 E-10 在浏览器中输入 localhost:8765/admin/userinfo 的结果

E.6.2 apollouser 服务运行结果

在浏览器中输入 localhost:8760/userinfo，结果如图 E-11 所示。在浏览器中输入 localhost:8760/guest/userinfo，结果如图 E-12 所示。

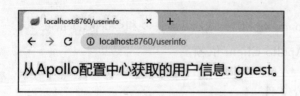

图 E-11 在浏览器中输入 localhost:8760/userinfo 的结果

图 E-12 在浏览器中输入 localhost:8760/guest/userinfo 的结果

E.6.3　zuulapollo 服务运行结果

在浏览器中输入 localhost:8665/admin/userinfo，结果如图 E-13 所示。在浏览器中输入 localhost:8665/admin/admin/userinfo，结果如图 E-14 所示。对比图 E-9 和图 E-13（或图 E-10 和图 E-14），可以发现它们结果相同，即 zuul 路由项目 zuulapollo 对 URL 进行了转换处理。

在浏览器中输入 localhost:8665/guest/userinfo，结果如图 E-15 所示。在浏览器中输入 localhost:8665/guest/guest/userinfo，结果如图 E-16 所示。对比图 E-11 和图 E-15（或图 E-12 和图 E-16），可以发现它们结果相同，即项目 zuulapollo 对 URL 进行了转换处理。

对比图 E-13 和图 E-15（或图 E-14 和图 E-16），可以发现项目 zuulapollo 对 URL 进行了解析并调用不同的服务（apolloconfig 或者 apollouser）为用户提供服务。

图 E-13　在浏览器中输入 localhost:8665/admin/userinfo 的结果

图 E-14　在浏览器中输入 localhost:8665/admin/admin/userinfo 的结果

图 E-15　在浏览器中输入 localhost:8665/guest/userinfo 的结果

图 E-16　在浏览器中输入 localhost:8665/guest/guest/userinfo 的结果

附录 F

视频讲解

Spring Cloud 在微信小程序的简单应用

可以以不同的方式访问 Spring Cloud 微服务，这使得 Spring Cloud 微服务可以作为微信小程序的后台。

本附录介绍以附录 E 中的 Spring Cloud 微服务作为微信小程序后台的一个简单应用。整个应用分为 Eureka 服务器项目 mweathereurekaserver、服务提供者项目 apolloconfig 和微信小程序项目 wxmpforsc。

本附录对 Spring Cloud 微服务与微信小程序的整合开发提供入门介绍。若要开发复杂应用还需要对 Spring Cloud 微服务开发和微信小程序开发有更深入的理解。

F.1 启动作为后台的 Spring Cloud 微服务

F.1.1 启动 Apollo 配置中心

在目录 apollo-builds-scripts-master 下启动 Git Bash，执行如例 E-1 所示的命令启动 Apollo 配置中心。保持附录 E 中 Apollo 配置中心默认项目 SampleApp 的配置内容不变。

F.1.2 保持后台服务不变

保持附录 E 中 Eureka 服务器项目 mweathereurekaserver、服务提供者项目 apolloconfig 的所有内容不变。具体代码请参考本书附带的源代码。

F.1.3 在浏览器中直接访问微服务的结果

依次在 IDEA 中运行项目 mweathereurekaserver 和 apolloconfig。在浏览器中输入 localhost:8765/userinfo,结果如图 F-1 所示。

图 F-1　在浏览器中输入 localhost:8765/userinfo 的结果

F.2　前端微信小程序应用的实现

F.2.1　微信小程序的开发工具安装和项目创建

为了帮助开发者简单、高效地开发微信小程序,微信官方推出了小程序开发工具。可以从官网下载该开发工具,并按照导航逐步安装。安装完成后,会在桌面上添加"微信开发者工具"图标。双击该图标打开"微信开发者工具",结果如图 F-2 所示。注意,开发者不同,图 F-2 中的二维码也会不同。

然后,用手机微信扫描二维码,通过验证后显示"扫描成功",结果如图 F-3 所示。显示该工具可以用来开发"微信小程序项目"和"公众号网页项目",结果如图 F-4 所示。选择"微信小程序项目"后可以创建一个微信小程序项目。微信开发者工具的安装和项目创建

图 F-2　打开"微信开发者工具"的结果

细节本书不作介绍，读者可以参考编者编写的《微信小程序开发基础》一书或其他资料。

图 F-3　通过验证后的显示"扫描成功"的结果

图 F-4　显示开发工具可以用来开发两类项目的结果

F.2.2　创建项目并新建、修改文件

创建微信小程序项目 wxmpforsc 后，依次在目录 pages 下添加 callspringcloudservice、userinfo 两组文件，每组包括 4 个文件（如 userinfo.js、userinfo.wxml、userinfo.wxss、userinfo.json 为一组文件）。增加 8 个文件后项目的目录、文件结构如图 F-5 所示，修改这些文件和项目根目录下的 app.json 文件。这些文件的具体代码请参考本书附带的源代码。

F.2.3　微信小程序项目的运行结果

完成编译微信小程序后，在 Nexus 6 模拟器中显示的首页界面（与 callspringcloudservice.wxml 文件对应）如图 F-6 所示。单击图 F-6 中"调用 apolloconfig 服务"按钮，跳转到如图 F-7 所示的界面（与 userinfo.wxml 文件对应）。此时微信开发工具控制台的输出内容是"从 Apollo 配置中心读取的用户信息：admin。"，对比此输出和图 F-1 中的输出，可以发现两者一致。

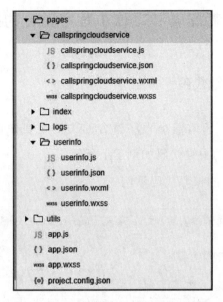

图 F-5　增加 8 个文件后项目 wxmpforsc 的目录和文件结构

图 F-6　项目 wxmpforsc 的首页界面

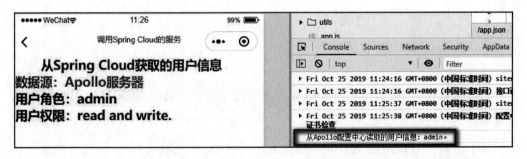

图 F-7　单击图 F-6 中"调用 apolloconfig 服务"按钮后跳转到的界面及控制台的输出内容

F.3　Spring Cloud 微服务和微信小程序整合的关键点

F.3.1　两者关联的关键代码

Spring Cloud 微服务和微信小程序整合的关键是在微信小程序中访问 Spring Cloud 微服务,两者关联的关键代码如例 F-1 所示。

【例 F-1】　两者关联的核心代码示例。

```
//下面源代码属于微信小程序项目中文件 pages/userinfo/userinfo.js
……//省略了代码
onLoad: function (options) {
    var that = this;
    wx.request({
      url: 'http://localhost:8765/userinfo',//微信小程序访问 Spring Cloud 微服务
      method: 'GET',
      data: {},
      success: function (res) {
        console.log((res.data));          //Spring Cloud 微服务返回给微信小程序的内容
         ……//省略了代码
      }
    })
  },
……//省略了代码
```

读者可以对比本案例与用 Spring Boot 应用作为微信小程序后台的情况(可参考编者编写的《Spring Boot 开发实战:微课视频版》一书或其他资料),可以发现 Spring Cloud 微服务作为微信小程序后台与 Spring Boot 应用作为微信小程序后台的关键点完全相同。

F.3.2　注意事项

在默认情况下,微信小程序和服务器进行网络通信的方式包括 HTTPS 协议(请注意不是 HTTP 协议)和 Websocket 等。开发时可以申请 HTTPS 域名;或者在开发工具中进行设置后使用 HTTP 域名(如 http://localhost:8765)。设置方法是单击工具中"详情"按钮,选择"本地设置"命令,选择"不校验合法域名、web-view(业务域名)、TLS 版本以及 HTTPS 证书"前的方框,如图 F-8 所示。

附录F　Spring Cloud 在微信小程序的简单应用

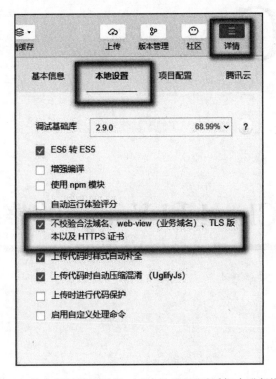

图 F-8　单击"详情"按钮后,在"本地设置"对话框中进行设置

附录 G

视频讲解

Spring Cloud 和 Vue.js 的整合开发

Vue.js 是构建用户界面的渐进式框架。Vue.js 采用自底向上增量开发的设计，较容易与其他库或已有项目整合。

本附录介绍附录 E 中的微服务整合 Vue.js（本附录使用的版本 2.6.10 版）的简单应用。整个应用分为 Eureka 服务器项目 mweathereurekaserver、服务提供者项目 apolloconfig 和 Vue.js 项目 helloworld。

本附录对 Spring Cloud 与 Vue.js 的整合开发提供入门介绍。若要开发复杂应用还需要对 Spring Cloud 微服务开发和 Vue.js 开发有更深入的理解。

G.1 在 IDEA 中整合 Spring Cloud 和 Vue.js

G.1.1 Vue.js 的安装

安装 Vue.js 之前先要安装 Node.js，到官网下载 Node.js 后，安装 Node.js。安装 Node.js 时就已经自带了包管理器 npm。安装完 Node.js 之后，打开 Windows 命令处理程序 CMD，依次执行如例 G-1 所示的一序列命令。

附录G　Spring Cloud 和 Vue.js 的整合开发

【例 G-1】　一序列命令的示例。

```
node -v
npm -v
npm install -g cnpm --registry=https://registry.npm.taobao.org
cnpm install -g vue-cli
cd d:
d:
md workspace
cd d:\workspace
vue init webpack  sb-vue
cd d:\workspace\sb-vue
npm run dev
```

例 G-1 中每一行代表一条命令，前一条命令执行完成之后，才执行后一条命令。如第 1 条命令执行完成之后才能执行第 2 条命令，结果如图 G-1 所示。第 3 条命令用来安装 cnpm，第 4 条命令用来安装 2.x 版脚手架，第 5 条至第 7 条命令在 D 盘新建目录 workspace（也可改成其他名字），第 9 条命令创建一个项目 sb-vue（也可改成其他名称），第 11 条命令运行项目。

图 G-1　在 Windows 命令处理程序 CMD 中执行完第 1 条命令后再执行第 2 条命令的结果

在执行第 9 条命令时，除了要给出作者的姓名（可以根据需要命名）之外，其余每项选择 Y 后单击回车键即可。

正确执行完例 G-1 的命令之后，在浏览器中输入 localhost:8080，结果如图 G-2 所示。

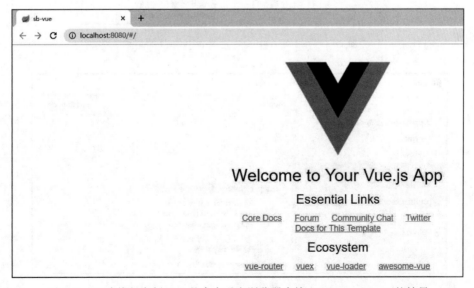

图 G-2　正确执行完例 G-1 的命令后在浏览器中输入 localhost:8080 的结果

G.1.2 在 IDEA 中集成 Vue.js

本书用的集成开发环境是 IDEA,安装完 Vue.js 之后可以将其集成到 IDEA 中。打开 IDEA 的 Plugins 窗口,在查询文本框中输入 Vue.js。单击搜索到的 Vue.js 插件图标和介绍文字下方的 Install 按钮,IDEA 自动安装 Vue.js 插件,安装结束后 Install 按钮变成 Restart IDE 按钮,结果如图 G-3 所示。单击 Restart IDE 按钮,重新启动 IDEA;再次打开 IDEA 的 Plugins 窗口并在查询文本框中输入 Vue.js,Install 按钮变成按钮 Installed 且按钮颜色变成灰色,显示已经成功安装 Vue.js 插件,结果如图 G-4 所示。打开 IDEA 后单击 Create New Project 并选择 Static Web 项目,可以看到已有 Vue.js 类型项目可供选择,如图 G-5 所示,说明成功地在 IDEA 中集成了 Vue.js。为了有效地使用 Vue.js,还要将 JavaScipt 的版本设置成 ECMAScript 6,如图 G-6 所示。

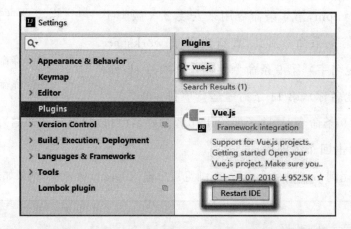

图 G-3　在 IDEA 中安装完 Vue.js 插件的结果

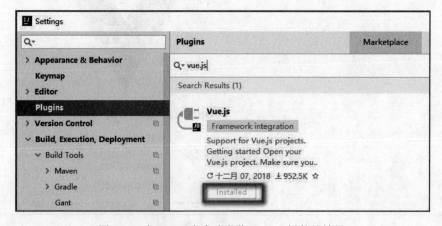

图 G-4　在 IDEA 中成功安装 Vue.js 插件的结果

附录G　Spring Cloud 和 Vue.js 的整合开发

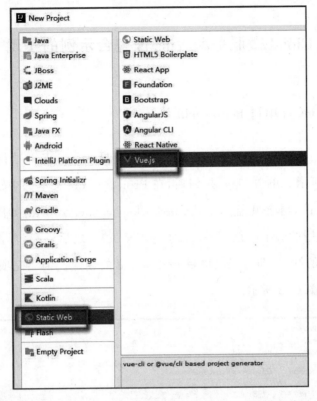

图 G-5　在 IDEA 中 Create New Project 并选择 Static Web 项目后可见到 Vue.js 项目的界面

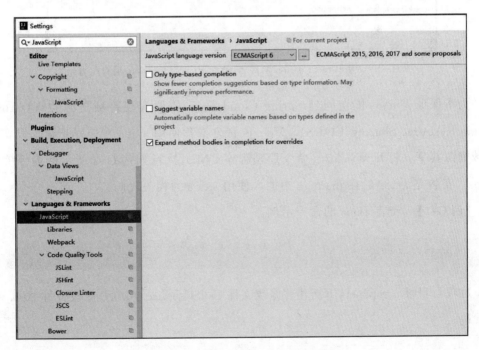

图 G-6　设置 JavaScipt 版本的界面

G.2　Spring Cloud 微服务和 Vue.js 整合示例的实现

G.2.1　创建 Vue.js 项目 helloworld

打开 IDEA 后,单击 Create New Project 并选择 Static Web 项目中 Vue.js 类型项目,如图 G-5 所示。单击 Next 按钮后,在 Project name 文本框中输入 helloworld,在 Project location 文本框中输入 D:\vue-workspace\helloworld,如图 G-7 所示。在图 G-7 中注意选用的脚手架是 vue-cli(对应 2.x 版)。之后,一直单击 Next 按钮(其中需要设置作者的姓名)即可完成项目 helloworld 的创建。创建完项目后,其主要目录和文件构成如图 G-8 所示。

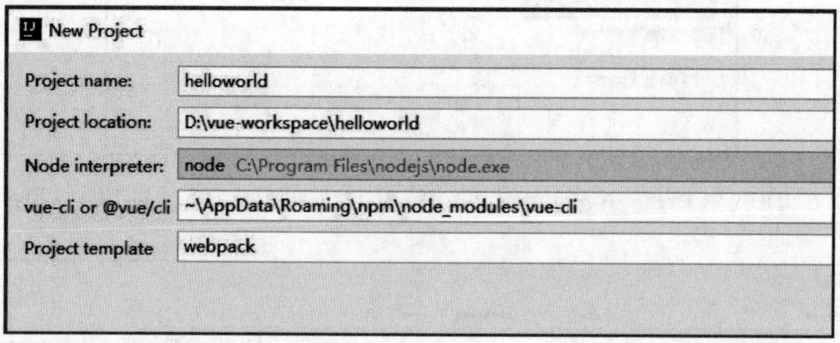

图 G-7　设置 Project name(项目名称)和 Project location(项目位置)

为了保证 Vue.js 能访问到 Spring Cloud 服务,需要实现跨域资源共享(Cross-Origin Resource Sharing,CORS)。Vue.js 官方推荐使用 axios 发送异步请求以实现跨域资源共享。打开 Windows 命令处理程序 CMD,执行如例 G-2 所示的命令安装 axios。安装完 axios 后在 main.js 中加入使用 axios 的相关代码。

【例 G-2】　安装 axios 的命令示例。

```
npm install axios -S
```

在项目目录 componets 下新建并修改文件 CallSCservice.vue 后,修改文件 App.vue 的代码。

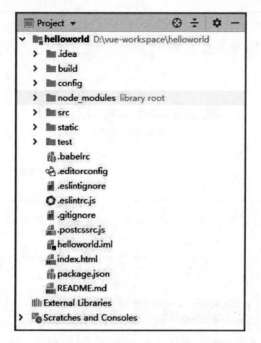

图 G-8　项目 helloworld 的主要目录和文件构成

G.2.2　后台服务

为了实现跨域资源共享，需要对项目 apolloconfig 中类 ACController 增加注解 @CrossOrigin。项目 apolloconfig 中其他内容保持不变，均和附录 E 相同。

保持 Eureka 服务器项目 mweathereurekaserver 所有内容不变，均和附录 E 相同。

上述文件的具体代码请参考本书附带的源代码。

G.2.3　运行结果

在 IDEA 中依次运行 Spring Cloud 项目 mweathereurekaserver、apolloconfig 和 Vue.js 项目 helloworld。在浏览器中输入 localhost:8082，结果如图 G-9 所示。单击图 G-9 中"调用 apolloconfig 服务"按钮，IDEA 控制台输出后台返回的数据（"从 Apollo 配置中心读取的用户信息：admin。"）如图 G-10 所示。对比图 G-10 和图 F-1 的输出以及图 F-7 中微信开发者工具控制台的输出，可以发现三者内容一致。

图 G-9　在浏览器中输入 localhost:8082 的结果

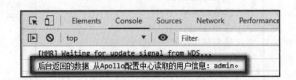

图 G-10　单击图 G-9 中"调用 apolloconfig 服务"按钮后 IDEA 控制台的输出

G.3　Spring Cloud 微服务和 Vue.js 整合的关键点

G.3.1　两者整合的关键

　　Spring Cloud 微服务和 Vue.js 整合的关键是实现跨域资源共享，即前端 Vue.js 项目能访问后台 Spring Cloud 微服务。为了实现这一目标，需要在 Spring Cloud 微服务中添加注解@CrossOrigin，以允许跨域访问。与此同时，Vue.js 项目使用 axios 来访问跨域资源（即微服务）。Vue.js 访问微服务的关键代码如例 G-1 所示。

　　【例 G-1】　Vue.js 访问微服务的关键代码示例。

```
// 源程序属于 Vue.js 项目中的文件 src/components/CallSCservice.vue
......//省略了代码
callscs:function() {
        this.$axios({
          url: 'http://localhost:8765/userinfo',//请求 Spring Cloud 微服务的完整地址
          method: 'get',                        //请求的方式
        }).then(res => {
```

```
            console.info('后台返回的数据', res.data); //Spring Cloud 微服务返回给
                                                     //Vue.js 项目的内容
        }).catch(err => {
            console.info('报错的信息', err.response.message);
        });
      }
    },
......//省略了代码
```

G.3.2 结果对比

对于相同的后台 Spring Cloud 微服务(项目 mweathereurekaserver、apolloconfig),以微信小程序为前端(附录 F)和以 Vue.js 为前端(附录 G)的访问方法相似,结果相同。

参考文献

[1] 郑天民. 微服务设计原理与架构[M]. 北京:人民邮电出版社,2018.
[2] 周立. Spring Cloud 与 Docker 微服务架构实战[M]. 2 版. 北京:电子工业出版社,2018.
[3] 柳伟卫. Spring Cloud 微服务架构开发实战[M]. 北京:北京大学出版社,2018.
[4] 许进,叶远志,钟尊发,等. 重新定义 Spring Cloud 实战[M]. 北京:机械工业出版社,2018.
[5] 方志朋. 深入理解 Spring Cloud 与微服务构建[M]. 北京:人民邮电出版社,2018.
[6] 尹吉欢. Spring Cloud 微服务:全栈技术与案例解析[M]. 北京:机械工业出版社,2018.
[7] 翟永超. Spring Cloud 微服务实战[M]. 北京:电子工业出版社,2017.
[8] Sam Newman. 微服务设计[M]. 崔力强,张骏,译. 北京:人民邮电出版社,2016.
[9] 黑马程序员. 微服务架构基础:Spring Boot+Spring Cloud+Docker[M]. 北京:人民邮电出版社,2018.
[10] Susan J Fowler. 生产微服务[M]. 薛命灯,译. 北京:电子工业出版社,2017.
[11] Flavio Junqueira,Benjamin Reedzhu. ZooKeeper:分布式过程协同技术详解[M]. 谢超周,贵卿,译. 北京:机械工业出版社,2016.